Failure of Fibre-Reinforced Polymer Composites

Failure of Fibre-Reinforced Polymer Composites

Edited by
Mohamed Thariq Hameed Sultan
Murugan Rajesh
Kandasamy Jayakrishna

CRC Press
Taylor & Francis Group
Boca Raton London New York

CRC Press is an imprint of the
Taylor & Francis Group, an **informa** business

First edition published 2022
by CRC Press
6000 Broken Sound Parkway NW, Suite 300, Boca Raton, FL 33487-2742

and by CRC Press
2 Park Square, Milton Park, Abingdon, Oxon, OX14 4RN

ISBN: 978-0-367-65315-6 (hbk)
ISBN: 978-0-367-65316-3 (pbk)
ISBN: 978-1-003-12886-1 (ebk)

DOI: 10.1201/9781003128861

Typeset in Times
by SPi Technologies India Pvt Ltd (Straive)

Contents

Illustration

FIGURES

TABLES

Preface

Composites are defined as materials made up of two or more different elements that, when combined, produce a tougher and more durable material than the individual elements. When composites were introduced in aircraft components and aerospace industries, unexpected impacts occurred. These may have been due to damage during flight operations, such as runway debris on composite airframes, bird strikes during flight operations and dropping of hand tools during maintenance. These damages could not be seen by the naked eye and were undetectable. Such undetected hidden damage is also known as barely visible impact damage (BVID). Matrix cracking, fibre fracture, fibre pull-out and delamination are major undetected hidden types of damage faced by composite materials in the event of a low-velocity impact. When these materials are subjected to low-velocity impacts, the structural integrity, stiffness and toughness of the material are significantly reduced, which in extreme scenarios can result in a catastrophic failure of the structure. There is therefore a need to study the behaviour of composite materials under impact loading, since impacts occur mainly during maintenance and manufacturing work.

This book provides insights into the failure of fibre-reinforced polymer composites, with a focus on fibre failure mode, fibre failure propagation and factors that affect fibre failures. The book covers a variety of topics, including natural fibre-reinforced polymer composites, environmental factors affecting the fibre–matrix interface, ageing and its influence on the mechanical properties of banana/sisal hybrid composites, and interfacial adhesion improvement of polymer composites using graphene fillers.

The editors would like to extend their sincere gratitude to all the chapter authors who have contributed their valuable findings, ideas and knowledge on the field of composites and composite coatings. We greatly appreciate their commitment and support.. Last but not least, we are most thankful to the CRC team for their generous cooperation at every stage of production of this book.

M.T.H. Sultan
Murugan Rajesh
Kandasamy Jayakrishna

Editor Biographies

Professor Mohamed Thariq Hameed Sultan completed his PhD in Mechanical Engineering at the University of Sheffield, UK in 2011. He specializes in the fields of hybrid composites, advanced materials, structural health monitoring and impact studies. He is a Professional Engineer (PEng) registered under the Board of Engineers Malaysia (BEM) and a Chartered Engineer (CEng) registered with the UK Institution of Mechanical Engineers (IMechE). He was recently awarded a PTech (Professional Technologist) by the Malaysian Board of Technologists (MBOT). He has published more than 220 journal articles and 15 books internationally.

Professor Murugan Rajesh completed his PhD in composite materials at the Department of Mechanical Engineering, National Institute of Technology Karnataka, Surathkal, Mangalore, India in 2017. He is currently working in the School of Mechanical Engineering at the Vellore Institute of Technology University, India. He specializes in the fields of hybrid composites, advanced materials, composite joints and their failure analysis. He has published more than 17 journal articles in leading SCI-indexed journals, as well as 17 book chapters internationally.

Dr Kandasamy Jayakrishna is an associate professor in the School of Mechanical Engineering at the Vellore Institute of Technology University, India. Dr Jayakrishna's research is focused on bio-composites and management of manufacturing systems and supply chains to enhance efficiency, productivity and sustainability. His more recent research is in the area of developing tools and techniques to enable value creation through sustainable manufacturing, including methods to facilitate more sustainable product design for closed-loop material flow in industrial symbiotic set-ups, and developing sustainable products using hybrid bio-composites. He has published 52 journal articles in leading SCI/SCOPUS-indexed journals, 23 book chapters, 90 refereed conference proceedings and one book in a CRC/Springer series. His initiatives to improve teaching effectiveness have been recognized by national awards. In 2019 he was awarded both the Young Engineer Award by the Institution of Engineers (India) and the Distinguished Researcher Award in the field of sustainable systems engineering by the International Institute of Organized Research.

Contributors

Furkan Ahmad
Department of Mechanical Engineering
Delhi Technological University
Shahbad Daulatpur, New Delhi, India

Takeshi Akinaga
Graduate School of Engineering Science
Akita University
Akita, Japan

S. Arulvel
School of Mechanical Engineering
Vellore Institute of Technology
Vellore, India

S. Arun
National Aerospace Laboratory
Bengaluru, India

Cagrihan Celebi
Department of Mechanical Engineering
Zonguldak Bulent Ecevit University
Zonguldak, Turkey

Vijay Chaudhary
Mechanical Engineering Department
Amity School of Engineering and Technology, Amity University
Noida, Uttar Pradesh, India

Thulasidhas Dhilipkumar
School of Mechanical Engineering
Vellore Institute of Technology
Vellore, India

D. Dsilva Winfred Rufuss
School of Mechanical Engineering
Vellore Institute of Technology
Vellore, India

B.N.V.S. Ganesh Gupta K.
Composite Materials Laboratory, Department of Metallurgical and Materials Engineering
National Institute of Technology
Rourkela, Odisha, India

FRP Composite Laboratory, Department of Metallurgical and Materials Engineering
National Institute of Technology
Rourkela, India

K. Jayakrishna
School of Mechanical Engineering
Vellore Institute of Technology
Vellore, India

Anita Jessie
School of Civil Engineering
Vellore Institute of Technology
Vellore, India

Deepak Kohli
Department of Chemical Engineering
National Institute of Technology
Jalandhar, Punjab, India

V. Lakshmi Narayanan
School of Mechanical Engineering
Vellore Institute of Technology
Vellore, India

Jose Machado
University of Minho, Portugal

Kishore Kumar Mahato
Department of Manufacturing Engineering, School of Mechanical Engineering
Vellore Institute of Technology
Vellore, Tamilnadu, India

Sabuj Mallik
Head of Engineering Discipline, College of Science and Engineering
University of Derby
Derby, United Kingdom

D. Mallikarjuna Reddy
Vellore Institute of Technology
Vellore, India

H. Manjunath
Siddaganga Institute of Technology
Tumkuru, Karnataka, India

K.M. Manjunatha Swamy
Siddaganga Institute of Technology
Tumkuru, Karnataka, India

Ankit Manral
MPAE Division
Netaji Subhas University of Technology (formerly NSIT)
Dwarka, New Delhi, India

M. Mohammed Yahaya Khan
Department of Mechanical and Manufacturing Engineering Technology
Jubail Industrial College
Kingdom of Saudi Arabia

Garje C. Mohan Kumar
Polymer Composites Laboratory, Dept. of Mechanical Engineering
National Institute of Technology Karnataka
Surathkal Srinivasanagar, Mangaluru, India

J. Naveen
Vellore Institute of Technology
Vellore, India

Priyanka
Siddaganga Institute of Technology
Tumkuru, Karnataka, India

Rajesh Kumar Prusty
FRP Composite Laboratory, Department of Metallurgical and Materials Engineering
National Institute of Technology
Rourkela, India

Murugan Rajesh
School of Mechanical Engineering
Vellore Institute of Technology
Vellore, India

Bankim Chandra Ray
Composite Materials Laboratory
Department of Metallurgical and Materials Engineering
National Institute of Technology
Rourkela, Odisha, India

FRP Composite Laboratory
Department of Metallurgical and Materials Engineering
National Institute of Technology
Rourkela, India

Oguzhan Sen
Department of Mechanical Engineering
Zonguldak Bulent Ecevit University
Zonguldak, Turkey

S. Senthilrajan
Department of Mechanical Engineering
A.V.C. College of Engineering
Mannampandal, Tamil Nadu, India

N. Shanmuga Priya
Siddaganga Institute of Technology
Tumkuru, Karnataka, India

Marappan Shanmugasundaram
Jazan University
Kingdom of Saudi Arabia

N. Shanmugavadivu
RVS College of Engineering and Technology
Coimbatore, Tamil Nadu, India

Shivasharanayya Swamy
REVA University
Bangalore, India

A. Soundhar
Department of Mechanical Engineering
Sri Venkateswara College of Engineering (SVCE)
Chennai, India

G. Venkatachalam
Vellore Institute of Technology
Chennai, India

N. Venkateshwaran
Department of Mechanical Engineering
Rajalakshmi Engineering College
Chennai, Tamilnadu India

T.V. Vineeth Kumar
Siddaganga Institute of Technology
Tumakuru, India

Mehmet Yetmez
Department of Mechanical Engineering
Zonguldak Bulent Ecevit University
Zonguldak, Turkey

1 Natural Fibre-Reinforced Polymer Composites

Newer Materials for Weight-Sensitive Applications

Garje C. Mohan Kumar
National Institute of Technology Karnataka, Mangaluru, India

Sabuj Mallik
University of Derby, Derby, United Kingdom

CONTENTS

1.1 INTRODUCTION

The engineering community has been seeking to reduce the weight of moving vehicles, space, sports and medical devices for several decades in order to improve comfort, and economy of operation and maintenance. New materials need to be developed principally to reduce costs as well as to improve the overall structure for transportability without compromising functionality. Improving performance, speed and payloads are also important in lightweight applications. Weight is a critical determinant of performance in most high-end automotive and space applications without compromising essential engineering requirements. In general, the use of composite materials is favoured in many lightweight applications. Mallik et al. (2011), for example, proposed the use of a lightweight Al/SiC metal matrix composite for automotive thermal management applications. Natural fibre-reinforced polymer composites offer the

DOI: 10.1201/9781003128861-1

advantages not only of lower material density but also of conservation of natural resources and reduced use of materials for manufacturing consumer and other goods.

Lightweight materials are also required in the building, packaging and agriculture sectors. The choice of material for lightweight structures needs to take account of significant loading and stress factors within the engineering design configurations. Lightweight materials are considered during the design, development and implementation stages for components and other resources, as are manufacturing and production capabilities at reasonable total costs. A successful development in lightweight materials has been the replacement of metallic parts by plastics, possibly for lower-load applications, and by plastics with additional reinforcement for engineering applications involving moderate and heavy loads.

Most applications use either natural fibres or man-made (synthetic) fibres. Natural fibres are materials that are obtained from plants or animals, and even sometimes from minerals. Natural fibres have played a significant part in the development of new weight-sensitive polymer composites. In general, such fibres are long hair-like materials that can be spun together into threads that behave like a continuous elongated phase in plastic-reinforced composites. These threads or ropes made with threads can be used as reinforcements in polymer composites. Natural fibres obtained from the processing of vegetables usually contain a form of lignin-enriched cellulose or glucose polymer of glucose.

Today, biocomposites are developed with a biodegradable matrix (Mohanty et al. 2001 and Joshi et al. 2004). The development of biocomposite lightweight materials has attracted greater interest in recent times, due to their biodegradability compared to other man-made fibres and composites.

1.2 NATURAL FIBRES AND THEIR ORIGIN

Natural fibres are classified according to their origin: cellulose derived from plants and vegetables; proteins from animals; and minerals. Plant and vegetable fibres are further classified according to the section of the plant from which they are derived: for example, seed fibre such as cotton; bast fibres from the major part of the stem; sisal fibres from leaves; linen from flax plants; and husk fibres such as coconut and areca. Fibres from animal tissues are categorized as long hairs such as wool, fur spreads such as angora, or continuous secretions such as silk fibres. Plant and vegetable fibres are also classified according to their wood or non-wood content. Wood fibres are further subdivided into softwood and hardwood, depending on their strength (Mohanty et al. 2005; John and Thomas 2008).

In composite manufacturing industries, fibres are usually referred to as wood based or agricultural crop based, depending on whether they are root, bast, leaf, seed or stem fibres. These fibres typically contribute significantly to structural performance and life. They provide successful reinforcement for plastic composites. Bast-based fibres are used in composites and are processed in various industries including textiles, paper, building materials, agro-fibre-based part composites and other similar industrial components. Natural fibre composites are further classified under three main categories: (a) natural fibres reinforced with a thermosetting resin for developing components for engineering applications; (b) natural fibres used as a filler in thermoplastics such as polypropylene for lightweight applications; and (c) high-strength composites, where longer natural fibres are used with suitable adhesives and compatibilizers to attain higher toughness in thermoplastics (Marston 2008). Wood-based fibres are classified based on the type and part of the plant from which they are derived. Bast fibres are found in the soft stem of many dicotyledonous plants.

The fibres may form clusters across several cells and in some cases the complete cluster is the fibre. Most of the fibres are firmly glued to neighbouring fibres by the natural pectic middle lamella. These fibres form quite strong, durable strands that can withstand the external chemical environment and processes like bleaching or other harsh fibre treatments. These plant fibres are flax, hemp, jute, ramie, etc.

Most hard-leaf fibres form smaller strands with short cells and are seen in monocotyledonous plants. These fibro-vascular bundles are composed of supportive strands primarily available in plant leaves. Hard-leaf fibres consist of xylem, phloem and other sheathing cells scattered in the sleek pithy matrix. The lignified cells are harder than the unprocessed soft fibres found in the dicots where the cellulose is associated mainly with pectic-based material.

1.3 THE STRUCTURE OF NATURAL FIBRES

Natural fibres usually consist of various amounts of lignocellulose, in which helical cellulose microfibrils lie within a matrix of lignin, hemicellulose or both. The matrix cellulose is a natural polymer built of glucose units, which gives structural strength and stability to the fibres and plant cell walls. Hemicellulose is a different polymer made of various polysaccharides, which may form a cementing matrix hydrogen bonded to the cellulosic microfibrils. Lignin is amorphous in nature, being a heterogeneous mixture of aromatic polymers and phenyl propane monomers. Lignin, being strongly hydrophobic, enhances the properties of the composite by acting as a coupling agent. This increases the stiffness of the cellulose/hemicellulose element in the composite.

The cell walls of a fibre do not form a homogeneous membrane. Every fibre has its own layer-by-layer structure consisting of a thin primary wall and a surrounding outer secondary cell wall. The thick secondary walls consist of three layers. It is the middle layer, consisting of a series of helical cellular microfibrils wound to form long-chain cellulose molecules, that provides the strength and mechanical properties of the fibres.

The great advantage of natural fibres used as reinforcement materials in lightweight composites is their recyclability and environmentally friendly parts and production processes. Table 1.1 shows sources of important fibres.

1.4 PROPERTIES OF NATURAL FIBRES

In the development of lightweight materials, natural fibre reinforcements substituted for synthetic materials can reduce the weight of the components by up to 40%, improving fuel efficiency in the automotive, transport and aerospace sectors, as well

TABLE 1.1
Fibre Sources and Origin

Source of Fibre	Species	Part/Origin
Abaca	Musa textiles	Leaf
Pineapple	Ananus comosus	Leaf
Sisal	Agave sisalana	Leaf
Flax	Linum Usita tissum	Stem
Hemp	Cannabis sativa	Stem
Jute	Corchorus capsularis	Stem
Kenaf	Hibiscus cannabinus	Stem
Ramie	Boehmeria nivea	Stem
Areca	Areca catechu	Seed
Cotton	Gossypium sp.	Seed
Coir	Cocos nucifera	Fruit

Source: Saravana Bavan and Mohan Kumar (2014).

TABLE 1.2
Properties of Natural and Synthetic Fibres

Fibre		Density (g/cm^3)	Tensile strength (MPa)	Young's modulus (GPa)	Elongation at break
Natural	Coir	1.15	131–175	4–6	15–40
	Jute	1.3–1.45	393–773	13–26.5	1.2–1.5
	Cotton	1.5–1.6	287–800	5.5–12.6	7.0–8.0
	Sisal	1.45	468–640	9.4–22.0	3.0–7.0
	Flax	1.50	345–1100	27.6	2.7–3.2
	Ramie	1.50	400–938	61.4–128	1.2–3.8
Synthetic	E-glass	2.5	2000–3500	70	2.5–3.0
	Carbon	1.7	4000	230–240	1.4–1.8

as appearance and comfort. The important point to note is that a considerable improvement in ductility and flexibility is achieved with partial substitution of natural fibres in the parts. Table 1.2 compares properties of selected natural and synthetic fibres.

The microfibrils in the fibre bundles form an angle with the fibre axis which is known as the microfibrillar angle. The characteristic values of the microfibrillar angles, which partly determine the extension or deformation characteristics of fibres, vary from fibre to fibre. A natural twist is accomplished by hemicellulose lignin elements in the amorphous matrix phase which is very complex. For example, hemp fibre has 6–8°, flax 6–10°, jute 6–7° and sisal 10–25°. Microfibrils vary in diameter from 10 to 25 nm and consist of a bundle containing 40–100 cellulose molecules in a chain. It is these structures that determine the mechanical strength of the fibre bundles.

1.5 USE OF NATURAL FIBRES IN COMPOSITES

Natural fibres are biodegradable and absorb more carbon dioxide than is used in their production or processing. From the environmental perspective, therefore, their use needs to be encouraged (Mohanty et al. 2005).

Saravana Bavan and Mohan Kumar (2011, 2013a,b) processed a lightweight composite polymer reinforced with natural fibres, focusing on fibres extracted from maize stalk, reinforced with general-use unsaturated polyester resin using the catalyst MEKP (methyl ethyl ketone peroxide) and cobalt octoate as an accelerator. The composite material was processed employing vacuum-assisted resin transfer moulding (VARTM) with varying fibre fractions. The natural and chemically treated fibres were characterized by the tensile test and Fourier transform infrared spectroscopy techniques. SEM (scanning electron microscopy) and EDAX (energy dispersive X-ray analysis) of the fibres were carried out. The results showed that the maize fibres reinforced with polymers had excellent tensile properties, which are needed for any composite materials. Saravana Bavan and Mohan Kumar (2010), in a detailed study of the availability, processing and potential use of natural fibres in composite materials, summarized the mechanical and physical properties of fibres and their applications in India.

Swamy et al. (2004) studied properties of fibres extracted from raw green areca and compared the strength of areca fibre with other known natural coir fibre. The areca fibres were chemically treated to study their effects. Areca fibre-reinforced composite laminates are prepared using a hot press. Phenol-formaldehyde laminates with different proportions of randomly distributed fibres have been developed by Mohan Kumar (2010). The compatibility of matrix resin and fibre areca reinforcement was studied using various tests of mechanical properties such as tensile and adhesion tests. Other tests for moisture absorption and biodegradability of areca-reinforced phenol-formaldehyde composite laminates were carried out and analyzed using the numerical finite element method.

Mohan Kumar et al. (2020) investigated the development of a matrix material poly(vinyl alcohol) cross-linked with glutaraldehyde suitable for natural fibre-reinforced composites for lightweight applications, and evaluated its dynamic mechanical properties. The composites were prepared using a conventional casting technique with the cross-linked polymer. Flexural properties for static conditions were studied using a three-point bending test. It was observed that the dynamic mechanical properties of the PVA were improved significantly for the higher glutaraldehyde cross-linking percentage in the rubbery plateau region. The cross-linked polymer had much greater storage and values of loss moduli, indicating a compact network structure among the matrix elements, as well as higher stiffness.

Nagamadhu et al. (2020) evaluated the dynamic mechanical properties of three different types of sisal fabric as a reinforcement material, and the water absorption effects of reinforcement with epoxy for lightweight composites. The effect of fabric thickness (grams per square metre) and weave on the textile properties of the fabric were studied. Composites using varying fractions of all types of sisal fabrics were processed and analyzed for their effect on the dynamic mechanical properties of the sisal composites. The effect of fibre thickness with multiple layers of reinforcement on the dynamic mechanical properties for frequency swap were also analyzed. The results showed that the temperature has a direct effect on the storage modulus, which decreases at elevated temperatures in all the sisal woven composites under study.

1.6 NATURAL FIBRE COMPOSITES FOR PACKAGING MATERIALS

The selection of polymers as matrix materials is crucial for developing materials for weight-sensitive application or parts. Selecting a polymer of lower density than the reinforcement but having comparable properties with other matrices, and without compromising the requirements, is always challenging. There are many examples of reinforced polymer composite with natural fibres instead of glass fibres used as a filler, which is particularly relevant for packaging and cushioning of agile parts for transport. More complex is the development of lightweight plastic cost-effective packaging for food preservation.

The choice of plastics over metal is supported by considerations of strength, weight and cost. Plastic is always preferred for short-life or disposable packaging, being more economical to manufacture, convenient to use and easy to dispose of. The disposal of plastics bags and containers is more expensive than their use, and their re-use or neutralizing needs to be managed.

On the other hand, some natural materials are incomparably superior to plastics. For example, wood-based packing material is uncompromised and appears acceptable from all points of view. Many studies on wood-based particleboards and plywood as alternatives to conventional wood are reported.

1.7 SHORT ARECA AND MAIZE FIBRE-REINFORCED COMPOSITES

Urea and phenol formaldehyde, a new matrix material as an alternative to conventional wood-based panels, began to be used in early 2000. Chemically processed natural fibres were reinforced with a controlled volume of urea and phenol formaldehyde. Published articles show that the composite panels or boards are inexpensive and the properties of the materials developed are more similar to wood-based panels. A number of studies on short areca and maize stalk fibre-reinforced composites are reported by Saravana Bavan and Mohan Kumar (2012, 2013a, and b, 2014).

Areca fibres are mechanically extracted from dried areca fruit husk by pulverizing, processed chemically and dried completely for enhancement of their surface properties. The mechanical properties of the extracted fibres and the effect of this treatment on fibre strength has been studied and compared with other known natural fibres such as coir. Areca fibre-reinforced composite laminates were prepared with randomly distributed fibres and phenol formaldehyde. Tensile, moisture absorption and biodegradable properties of these composites were tested and the areca-reinforced phenol formaldehyde composite laminates were shown to be suitable for weight-sensitive applications. Fibres extracted from areca fruit appear promising materials for reinforcing materials compared to all other natural fibres because the areca is from a potential perennial crop and the husk is less expensive as it is waste produced during the curing of areca. The areca, or *Areca catechu L.*, belongs to the family **Arecaceae** species originating in the Malay Peninsula. The areca husk or

TABLE 1.3
Selected Natural Fibre Composition Compared

Fibre	Lignin %	Cellulose %	Hemicellulose %
Coir	40–45	32–43	0.2–0.3
Areca	12–25	1–5	35–65
Sisal	10–14	66–72	12–15
Maize stalk	10–13	38–42	21–23
Banana	5–8	63–64	19–21
Flax	2–5	72–80	20–22
Cotton	1–3	78–85	5–7

shell is a long hard fibrous portion enclosing the endosperm and constitutes about 40% of the entire volume of the areca fruit. Areca husk fibres are mainly composed of hemicelluloses and less cellulose. Table 1.3 compares the composition of areca fibres and other known fibres.

The inner portion of the husk is irregularly lignified hard cells, and the middle layer contains soft fibres. Areca fibre is highly hemi-cellulosic and is larger than any other fibres. Coir fibre has higher lignin content than other fibres, while sisal fibre has a higher cellulosic content than fibres extracted from banana, sisal and other plants. Maize stalk fibres have considerable lignin and hemicelluloses compared to the cellulosic content of areca fibres. The properties of natural fibres vary depending on the locality in which they have grown, the age of the plant, and the fibre extraction method.

More than 100 areca fibre samples were prepared and their mechanical properties, such as tensile strength, Young's modulus and elongation at break were studied by Swamy et al. (2004). The stress-strain diagrams for the different areca fibres from the husk results show that the ultimate stress is around 85–102 MPa with 10–12% elongation at break. The fracture surface of fibres is ductile. The Young's modulus varies in the range of 1.1–12 GPa. The tensile experiment conducted on coir for the comparison shows that its ultimate stress is around 110–140 MPa with 10–12% elongation at break and the Young's modulus is 2.5–3.6 GPa. The ultimate strength and elongation at break of areca fibre are slightly lower than those of coir fibre.

Dynamic mechanical analysis of polymer composites of areca fibre reinforced with general-purpose unsaturated polyester were carried out by Mohan Kumar et al. (2008) at the Centre for Composite Manufacturing at the University of Delaware, USA.

In this work, the mechanical strength and dynamic properties of natural areca fibre reinforced in general-purpose unsaturated polyester are studied. Areca fibre–polymer composites were prepared using the VARTM with randomly distributed fibres. Storage and loss modulus of these composites were determined using a dynamic mechanical analyzer for visco-elastic properties, and flexural strength of the composites was determined and reported. Analysis of mechanical strength and dynamic mechanical analysis shows that short areca fibres are very good, promising fibres for general engineering applications.

Areca fibre-reinforced polymer composites were prepared using the vacuum-assisted resin transform method with randomly distributed fibres. Different composites were prepared with different proportions of areca fibre and polymer resin: 5, 10, 15, 20 and 25% of fibres in the polymer resin. An acrylic mould was prepared to cast a 100 mm × 100 mm × 3 mm composite. Areca fibres were distributed uniformly in the mould, and then the air was entirely evacuated within the vacuum bag. Later resin was infused into the mould to obtain a thick uniform composite. Figure 1.1 shows the casting of a composite by VARTM. After cure, the cast composites were taken out and post-cured at elevated temperature in the oven. Different composites with increasing fibre content were prepared.

Areca–polymer composite specimens were prepared for mechanical dynamic analysis as per ASTM D5023 and for bending as the standard ASTM D790 from the cast composites at the Center for Composite Manufacturing, University of Delaware. The loss and storage moduli and glass transition temperature T_g of the polymer were determined using a dynamic mechanical analyzer (DMA 2980, TA Instruments) at a heating rate of 5 °C/min according to ASTM D5023. Figure 1.2 shows the variation of loss and storage modulus and phase angle (Tan δ) for glass transition temperature. The T_g was obtained from the maximum value of loss factor tan δ. These composites exhibit closer values of storage modulus and glass transition temperature. The flexural strength of areca–polymer composites was determined according to ASTM D790 at room temperature using the Instron 4201 universal testing machine. The specimen was subjected to bending at recommended speed until failure occurred. During the test, the load and deflections for each specimen were recorded. The flexural stress and strain were calculated for each load case. Figure 1.3 shows the variation of these stresses and strains for different composition of fibre and areca–polymer composite matrices. The bending stresses in the areca–polymer composites vary significantly with the fibre propositions (Figure 1.3). For the composite AS05 with 5% areca fibre, MPa is 20.8 MPa, a figure that increases to 31.27 MPa for the composite AS25 of which areca is 25%. The rise in the specific strength of these composites is up to 52%. The Young's modulus for AS05 is 414 MPa and for AS25626 it is MPa. The graph shows that the maximum bending stresses and Young's modulus of the composite material increase with the increase in fibre content. The fibre loading in

FIGURE 1.1 Polymer composite processing by VARTM.

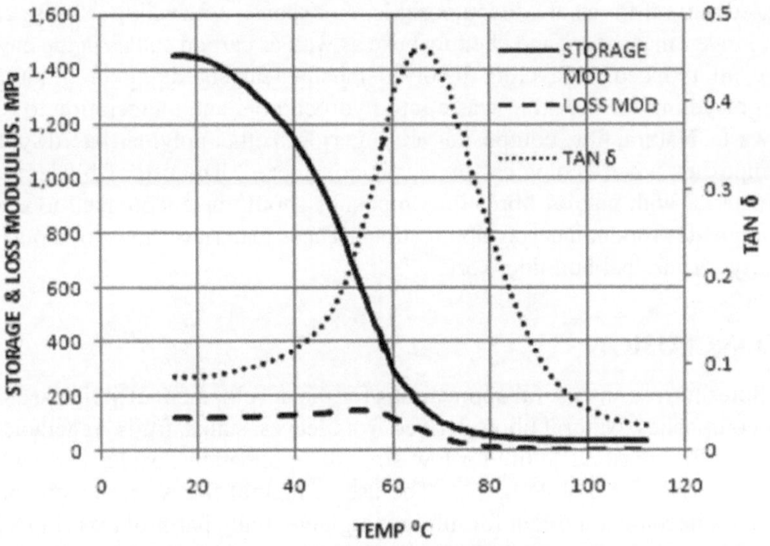

FIGURE 1.2 Dynamic mechanical analysis of areca–polymer composites.

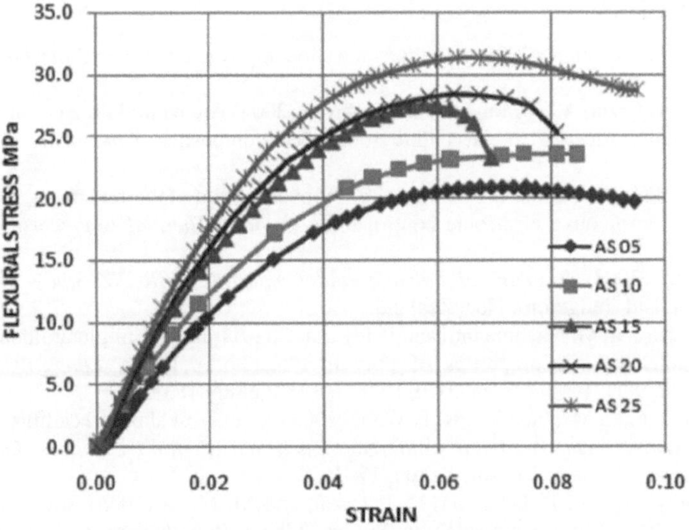

FIGURE 1.3 Variation of flexural stresses and strains for areca–polymer composite.

the composites will increase mechanical strength while the density of the composites decreases considerably. A composite for lightweight applications can be developed with natural fibre-reinforced composites.

The sustainability and ecological impacts of natural fibre-reinforced polymer composites for lower-weight applications need to be studied. To keep the environment safe and clean these composites need to be biodegradable or recylable.

Agro-based natural fibres are decomposable, so expensive recycling process can be avoided. Power, man-hours and chemical use as well as carbon values in the environment are all reduced. Recycling involves mechanical shredding or granulation, chemical recycling for breaking waste into hydrocarbon, and incineration to reduce landfill waste. Natural fibre composites are superior to other polymer-based synthetic fibre composites when ecological impact is considered. The use of biodegradable matrix material with natural fibres in composites should be encouraged to develop green composites for engineering applications. These materials show good potential, particularly for internal building work.

1.8 CONCLUSION

Natural fibres have many useful applications for the development of weight-sensitive polymer composites. Natural fibres derived from leaves, stalks, fruits, vegetables and animals provide a reasonable life for low-strength applications, packaging, and construction materials for mobile homes. The use of natural fibres as precious natural resources has become important for minimizing potentially harmful toxic emissions, and should be practised in the design and manufacture of items ranging from small household appliances to large jet planes and rockets.

REFERENCES

M. J. John and S. Thomas (2008) Biofibers and biocomposites, *Carbohydrate Polymers*, 71, 343–364.

S. V. Joshi, L. T. Drzal, A. K. Mohanty, and S. Arora (2004) Are natural fibre composites environmentally superior to glass fibre reinforced composites, *Composites Part A*, 35, 371–376.

S. Mallik, N. Ekere, C. Best, and R. Bhatti (2011) Investigation of thermal management materials for automotive electronic control units, *Applied Thermal Engineering*, 31(2–3), 355–362.

N. J. Marston (2008) *Bio-derived Polymer and Composites, BRANZ Study Report 192*, BRANZ Ltd, Judgeford, New Zealand.

G. C. Mohan Kumar, M. Nagamadhu, and P. Jeyaraj (2020) Influence of glutaraldehyde cross-linker on dynamic properties of polyvinyl alcohol polymer, *Emerging Materials Research*, 9(1), 1–13, March 2020, DOI:10.1680/jemmr.18.00059

G. C. Mohan Kumar, M. Zhan and R. P. Wool (2008) A study of short areca fibre reinforced PF composites, *International Conference on Materials for the Future*, Government Engineering College, Trissur, Kerala, 17–19

A. K. Mohanty, W. Liu, P. Tummala, L. T. Drzal, and M. Misra (2005) Soy protein-based plastics, blends and composites, in *Natural Fibres, Biopolymers, and Biocomposites*, (Eds). Mohanty, A. K., Misra, M., and Drzal, L. T., CRC Press, Taylor & Francis Group, Boca Raton, FL, pp. 699–725.

A. K. Mohanty, M. Misra, and G. Hinrichsen (2001) Biofibers, biodegradable polymers and biocomposites: an overview, *Macromolecular Materials and Engineering*, 276–277, 1–24.

M. Nagamadhu, P. Jeyaraj, G. C. Mohan Kumar (2020) Influence of textile properties on dynamic mechanical and thermal analysis of epoxy composite reinforced with woven sisal fabrics, *Sadhana* 45(1), 1–10, DOI:10.1007/s12046-019-1249-zS

D. Saravana Bavan and G. C. Mohan Kumar (2010) Potential use of natural fibre composite materials in India, *Journal of Reinforced Plastics and Composites*, 29(24) 3600–3613, DOI: 10.1177/0731684410381151

D. Saravana Bavan and G. C. Mohan Kumar (2011) Examination of tensile strength and FTIR features of maize fibres reinforced polymer composites, *International Journal on Recent Trends in Engineering & Technology*, 05(04), DOI: 01.IJRTET.05.04.190

D. Saravana Bavan and G. C. Mohan Kumar (2012) Chapter 4: maize-natural fiber as reinforcement with polymers for structural applications. *Biopolymers, Biomaterials, and Their Composites, Blends, and IPNs*, Volume-2, Recent Advances in Materials Sciences book series, Apple Academic Press Inc., Palm Bay, FL.

D. Saravana Bavan and G. C. Mohan Kumar (2013a) Finite element analysis of a natural fibre (maize) composite beam, *Journal of Engineering*, 2013, 1–7, DOI: 10.1155/2013/450381

D. Saravana Bavan and G. C. Mohan Kumar (2013b) Chapter 7: re-use of natural plant fibers for composite industrial applications, *Recycling and Reuse of Materials and Their Products*, Advances in Materials Science, (Eds). Grohens, Y., Kumar, S. K., Boudenne, A., Apple Academic Press, Burlington, Canada, January 2013, ISBN: 978-1-926895-27-7.

D. Saravana Bavan and G. C. Mohan Kumar (2014) Tensile and thermal degradation characteristics of vetiver fibre composites, *Proceedia Material Science*, 5, 611–615.

R. P. Swamy, G. C. Mohan Kumar, and Y. Vrushabhendrappa (2004) Study of areca-reinforced phenol-formaldehyde composites, *Journal of Reinforced Plastics and Composites*, 23(13), 1373–1382, DOI:10.1177/0731684404037049

2 Ageing and Its Influence on Mechanical Properties of Banana/Sisal Hybrid Composites

An Experimental and Analytical Approach

S. Senthilrajan
A.V.C. College of Engineering, Mannampandal, Tamil Nadu, India

N. Venkateshwaran
Rajalakshmi Engineering College, Chennai, Tamil Nadu, India

M. Mohammed Yahaya Khan
Jubail Industrial College, Kingdom of Saudi Arabia

CONTENTS

DOI: 10.1201/9781003128861-2

2.1 INTRODUCTION

Nowadays, the term "eco-materials" is used to describe materials that help to protect the atmosphere. Plant fibre-reinforced composites have stronger thermal and mechanical properties. Natural fibres have been used in preference to synthetic fibres for automotive parts, aerospace applications, and construction materials. Their high specific strength and modulus, and non-abrasive nature, as well as availability and strength, light weight, low cost, biodegradability, and fewer health hazards, are considered the main advantages of natural fibre-reinforced composites [1]. Barath et al. [2] found that the composites have a greater resistance to water absorption than wood-based particle boards, which have a water absorption rate of more than 40%. Results of tests on areca-reinforced UF composite plates show that this is a promising material for packaging and other general structural applications with moderate toughness. For example, areca fibre and maize powder absorb much less moisture than wood-based particle board. Pothen [3] studied the mechanical properties of short banana fibre and glass fibre, including tensile strength, impact strength, stress relief, and water absorption behaviour. The tensile strength and impact strength of the sample improved as the glass fibre content increased. Jawaid et al. [4] examined composites of empty oil palm fruit bunches and jute fibres, and their water absorption, thickness swelling, and density behaviour, concluding that EFB composites had the highest thickness swelling properties among the various types of composites. Pure jute composites absorbed the least amount of water and swelled the least. Pure jute fibre composites have the highest density, according to tests of their physical properties. The dimensional stability of oil palm EFB composites can be improved by hybridizing them with jute fibres. The influence of moisture on the mechanical properties of hemp fibre-reinforced unsaturated polyester composites was analyzed by Dhakal et al. [5] They found that moisture absorption increases with the fibre volume fraction, due to increased void sand cellulose content in the sample. Water uptake behaviour is radically altered at elevated temperatures due to significant moisture-induced degradation. Exposure to moisture results in significant drops in tensile and flexural properties due to the degradation of the fiber–matrix interface. Kalaprasad et al. [6] investigated the tensile properties of short sisal fibre-reinforced LDPE with various fibre loadings and compared the results to existing reinforcement theories. They concluded that the tensile properties of short fibre-reinforced composites are influenced by fibre length, fibre loading, fibre dispersion, fibre orientation, and fibre–matrix interfacial bond strength in all theoretical models. Sreekala et al. [7] used phenol formaldehyde as a matrix to investigate the effect of mechanical properties of oil palm fibre with glass fibre. Based on their findings, they concluded that using 40 wt% fibre as reinforcement results in better mechanical behaviour. Venkateshwaran et al. [8] studied a banana/sisal hybrid composite's mechanical and water absorption properties. Mechanical properties improved as fibre loading increased up to 50%, while water absorption decreased. Khan et al. [9] investigated the impact of stacking structure on the mechanical properties of kenaf/jute/epoxy hybrid composites. The findings show that composites made of kenaf/jute/kenaf (K/J/K) have stronger mechanical properties than jute/kenaf/jute (J/K/J) composites. They also concluded that K/J/K composites had good mechanical properties due to improved fibre–matrix adhesion. Mahesha et al. [10] studied *Grewia serrulata* bast fibre reinforced with polyester resin. For surface modification of the fibre, they used alkali, permanganate, acetylation, and silane treatments. The chemically treated composites had significantly better behaviours than the untreated ones. Saw et al. [11] analyzed the water absorption of a jute/coir hybrid composite, together with its swelling, density, and morphology behaviours They found that the hybridization of jute fibre with coir composites enhances the dimension stability and extensibility, and improves its water absorption, swelling and mechanical properties. Dong [12] studied the mechanical behaviour of natural fibre–hybrid composites, using banana, sisal, jute, flax, and hemp. The Mori–Tanaka method was used to determine the material behavior of the composites. They concluded that the skin lamina had lower stiffness and the core lamina higher stiffness, which was used as the tensile face of the composite.

The research reported in the present chapter studied the effect of the layering sequence with unidirectional fibre orientation on the mechanical properties of banana/sisal hybrid composites. Comparisons were made with the tensile properties of banana/epoxy, sisal/epoxy and hybrid of banana/sisal/epoxy composite, using parallel and series models, and Hirsch's model with experimental values.

2.2 MATERIALS AND METHODS

The composite laminate was made by hand using a lay-up technique. The mould used was 300 mm × 300 mm × 3 mm in size. Epoxy LY556 mixed with hardener HY951 was used to fabricate the composite plate. The weight ratio of epoxy to hardener was 10:1. Resin was purchased from local resources. The test samples were taken out to the required sizes according to the ASTM standard. Table 2.2 and Figure 2.2 show the different types of composite fabricated. Individual fibre modulus and strength values were found by UTM (Figure 2.1). Properties of the fibre and matrix are shown in Table 2.1.

TABLE 2.1
Properties of Natural Fibres and Resin

S. NO	Properties	Banana fibre (actual value)	Literature value	Sisal fibre (actual value)	Literature values	Epoxy resin	Literature values	References
1	Diameter (μm)	315–415	80–250	118–248	200	2	—	[13]
2	Tensile strength (Mpa)	425.6	529–759	526.7	350	10	35–100	[13, 14]
3	Tensile modulus (Gpa)	18.65	8.2	13.29	12.8	0.3	3–6	[13, 14]

FIGURE 2.1 (a) Banana fibre (b) sisal fibre (c) experimental set-up for individual fibre.

TABLE 2.2
Different Types of Composite

S.NO	Type of composite	Symbol
1	BANANA, BANANA, BANANA	BBB
2	SISAL, SISAL, SISAL	SSS
3	BANANA, SISAL, BANANA	BSB
4	SISAL, BANANA, SISAL	SBS

2.3 TESTING STANDARDS

The specimens were prepared for various mechanical tests in accordance with the ASTM standard. An ASTM D3039 [15] tensile test was carried out at a test speed of 5mm/min. by UTM. The flexural strength was found using the ASTM D 790-07 [16] procedure at a test speed of 5mm/min. The tensile and flexural tests were carried out using an H10K Tinius Olsen Universal Testing Machine. The impact strength was determined using an Izod impact tester as per ASTM standard D256 [17]. In each case five specimens were tested to obtain the average value. The water absorption test was done according to ASTM standard D570-98 [18].

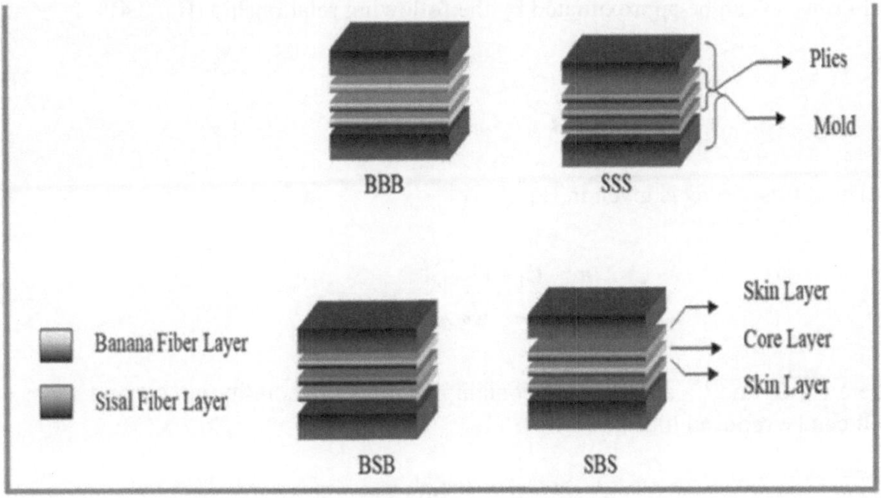

FIGURE 2.2 Arrangement of composite layers in unidirectional orientation.

2.4 WATER ABSORPTION BEHAVIOUR

According to ASTM D570, the specimen is placed in a container of distilled water maintained at a room temperature. and should be entirely immersed. At the end of 24 h, the specimens are removed from the water one at a time, all surface water is wiped off with a dry cloth and the specimens immediately weighed to the nearest 0.001g. The percentage of water absorption is calculated as:

$$\% \text{ of water absorption} = \frac{\mathbf{M_2 - M_1}}{\mathbf{M_1}} \tag{2.1}$$

where M_1 and M_2 are weight of dry and wet samples.

A diffusion test consists of measuring weight gain as a function of time when the specimen is exposed to a constant temperature and humidity environment. The diffusivity D is calculated from the initial linear relationship (Eq. 2.2) [19].

$$G = 1 - \frac{8}{\pi^2}\sum_{j=0}^{\alpha}\frac{1}{(2j+1)^2}exp\left[-\frac{(2j+1)^2\pi^2 Dt}{S^2}\right] \tag{2.2}$$

It can be approximated as Eq. (2.3)

$$G = 1 - \frac{8}{\pi^2}exp\left(-\frac{\pi^2 Dt}{S^2}\right) \tag{2.3}$$

where

s = h if material is exposed on both sides and

s = 2h if the material is exposed on one side only.

Further, it can be approximated by the following relationship (Eq. 2.4):

$$G = \frac{C - C_0}{C_m - C_0} = 4\left(\frac{Dt}{\pi S^2}\right)^{1/2} \tag{2.4}$$

The diffusivity D is given in Eq. (2.5)

$$D = \frac{\pi}{16}\left(\frac{C_1 - C_2}{C_m - C_0}\right)^2\left(\frac{S}{\sqrt{t_1} - \sqrt{t_2}}\right)^2 \tag{2.5}$$

where C_1, C_2 and C_m are moisture content and moisture equilibrium content.

It can be reduced to Eq. (2.6)

$$D = \pi\left(\frac{\mathbf{h\theta}}{\mathbf{4Q_\infty}}\right) \tag{2.6}$$

where

Θ is the slope of the linear portion of the sorption curve and

h is the initial sample thickness in mm.

The ability of solvent molecules to move among polymer segments is known as coefficient of diffusion. The absorption coefficient of water by the fibre is determined by permeability of water molecules. The sorption coefficient is determined from Eq. (2.7)

$$\mathbf{S} = \frac{\mathbf{Q_\infty}}{\mathbf{Q_t}} \tag{2.7}$$

The coefficient of permeability P (mm^2/s) is used to determine the effect of sorption and diffusion is written as in Eq. (2.8)

$$\mathbf{P = D X S} \tag{2.8}$$

2.5 REINFORCEMENT MODELS

In this study, individual sisal fibre and banana fibre were obtained to make 20 specimens, then a tensile test was carried out by UTM for 35 mm gauge length specimens. The diameter was found using an optical microscope. The tensile strength and tensile modulus values were substituted in the reinforcement models of the various theories, including parallel and series model, Hirsch's model, Einstein and Guth equations, Kerner equation, Cox model, etc. In this work the parallel and series model, and Hirsch's model are used.

2.5.1 Parallel and Series Model

This type of reinforcement model helps to find the Young's modulus and tensile strength. In the parallel model it is assumed that isostrain conditions exist for both matrix and fibre, whereas in the series model, stress is assumed to be uniform in both matrix and fibre [6, 7, 19]. The two models are shown in Equations (2.9)–(2.12).

Parallel Model

$$\mathbf{M_c = M_f V_f + M_m\, V_M} \tag{2.9}$$

$$\mathbf{TC = T_f V_f + T_m V_m} \tag{2.10}$$

Series Model

$$\mathbf{M_c \frac{M_{mM_f}}{M_{m\,V_f} + M_f V_m}} \tag{2.11}$$

$$\mathbf{T_c = \frac{T_f T_m}{T_m V_f + T_f V_m}} \tag{2.12}$$

2.5.2 Hirsch's Model

This model incorporates parallel and series components. Young's modulus and tensile strength are determined using the formulae in Eqs 2.13–2.14. In this equation x is a parameter used to determine the agreement between the experimental and reinforcement model. A control factor is also assumed [6, 7, 19].

$$\mathbf{M_c = X(M_m V_m + M_f V_f) + (1 - x)\frac{M_f M_m}{M_m V_f + M_f V_m}} \tag{2.13}$$

$$\mathbf{T_c = X(T_m V_m + T_f V_f) + (1 - x)\frac{T_f T_m}{T_m V_f + T_f V_m}} \tag{2.14}$$

where

$\mathbf{M_c}$ = Tensile modulus **Tc** = Tensile strength
$\mathbf{M_f}$ = Tensile modulus of fibre $\mathbf{V_f}$ = Fiber volume fraction
$\mathbf{M_m}$ = Tensile modulus of matrix $\mathbf{V_m}$ = Matrix volume fraction
$\mathbf{T_m}$ = Tensile strength of matrix $\mathbf{T_f}$ = Tensile strength of fibre

2.6 RESULTS AND DISCUSSION

2.6.1 Fundamental Water Absorption Mechanism

Natural fibres are often hydrophilic in nature. Water absorption capacity, capillarity and percentage of natural fibres is very high compared with synthetic fibres. Polymer matrixes are hydrophobic. Hence in the composite, the reinforcement of fibres and matrix can enhance the low adhesion between them. This low adhesion can cause

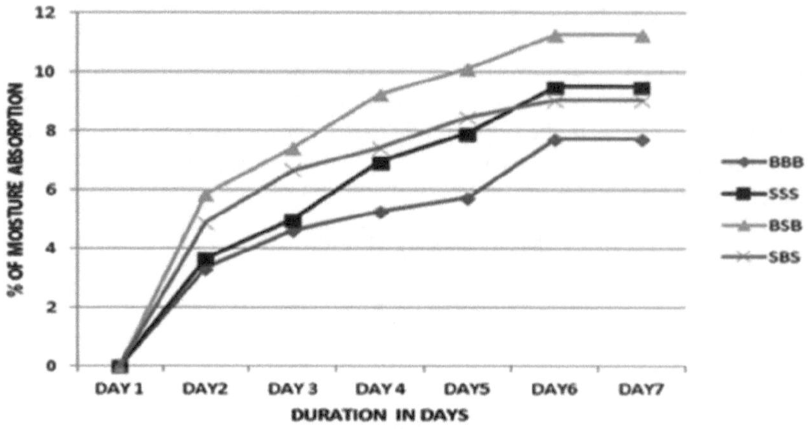

FIGURE 2.3 Moisture absorption vs duration for all types.

chemical affinity and very low polarity between fibre and matrix. These types of properties reduce the performance of the composite due to formation of voids and failure [10]. Permeability, temperature, volume fraction, orientation of the fibres, porosity, surface area and its protection, and diffusivity are associated with the intensity of water absorption [20–22].

In the composite, absorption by fibres occurs due to dissolution of water, which is then transported and deposited at the surface, which can cause micro cracks on the surface area of the composite. The water absorption mechanism consists of (i) diffusion (ii) capillarity and (iii) transportation. Moisture absorption occurs due to the random molecular motion of water from higher concentration to lower concentration areas. Figure 2.3 explains the water absorption mechanism of natural fibre composites. Water absorption induces swelling, plasticization, and a weakening of the composite's interfacial strength. In this moisture absorption phenomenon, the natural fibres have a cellulose content which contains hydrogen bonds or hydroxyl groups (OH). These hydroxyl groups exert a major influence on moisture absorption of banana fibre and sisal fibre. Water absorption increases as the fibre content of banana and sisal increases [23]. The cellulose content of banana and sisal fibres increases the penetration of water molecules to the interface of the composite through micro cracks [24]. Figure 2.4 shows the water absorption mechanism (Table 2.3).

Figure 2.5 shows moisture absorption vs duration for all types of composites. The regression model technique shown in Figure 2.5 is used to analyze the diffusion coefficients of BBB, SSS, BSB and SBS. There is no significant difference between the experimental and regression models. This linear regression curve correlates with our data. The value of the regression curve is 0.944, which shows the accuracy of experimental and theoretical models [25].

2.6.2 Tensile Properties of Composites

Figures 2.6 and 2.7 show the average tensile modulus and tensile strength of dry and wet specimens. The tensile properties of BSB are 102.32 MPa and 3.37 GPa

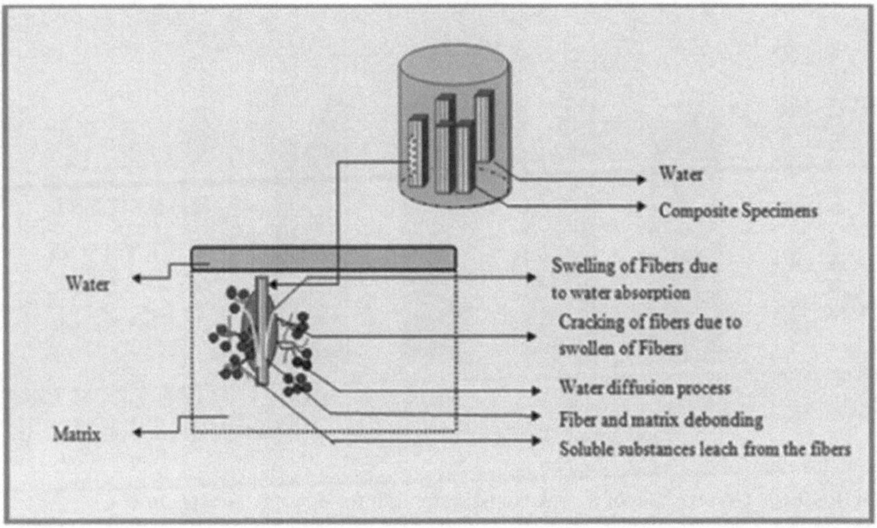

FIGURE 2.4 Moisture absorption mechanism.

TABLE 2.3
Moisture Absorption Behaviour of Banana, Sisal and Hybrid Composite

S. NO	Composite	Coefficient of diffusion (D) (mm^2/s)	Coefficient of sorption (S)	Coefficient of permeability (P) (mm^2/s)
1	BBB	0.370143	2.318	$8.348e^{-1}$
2	SSS	0.361951	2.595	$9.911e^{-1}$
3	BSB	0.351545	1.927	$6.766e^{-1}$
4	SBS	0.344808	1.840	$6.376e^{-1}$

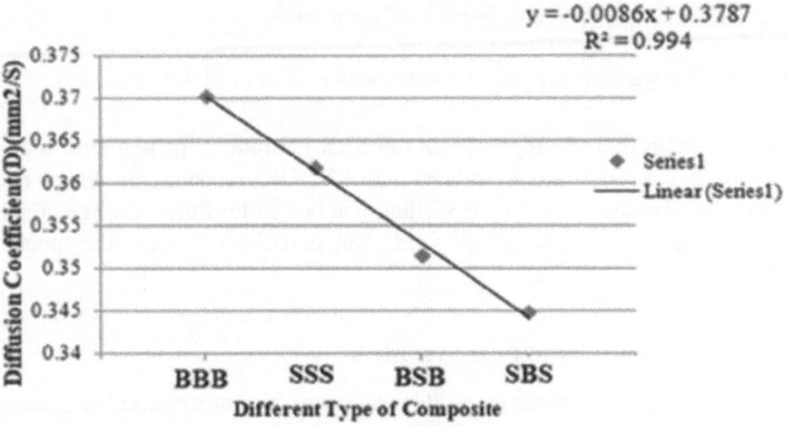

FIGURE 2.5 Regression model for diffusion coefficient.

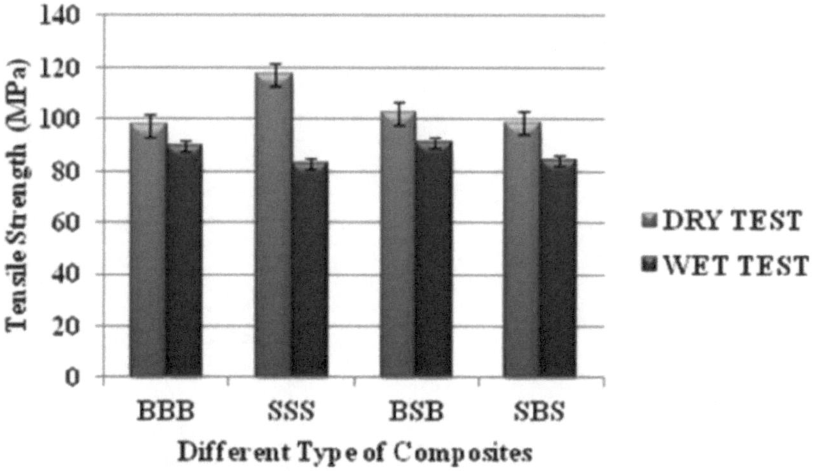

FIGURE 2.6 Comparison of average tensile strength for dry and wet specimens.

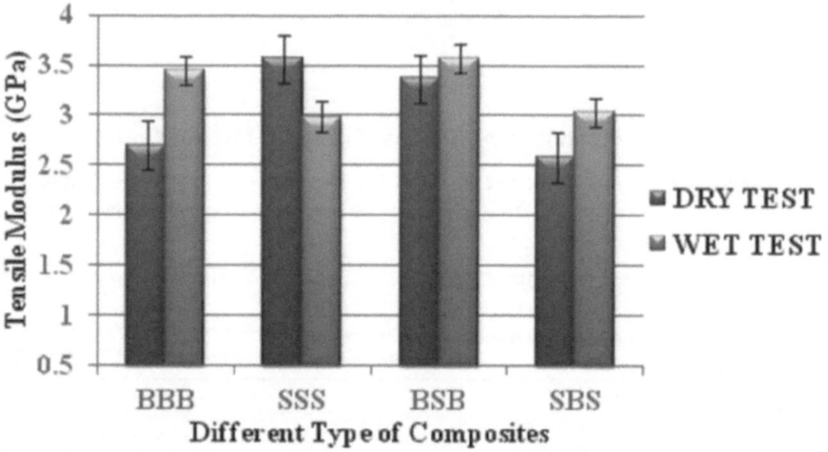

FIGURE 2.7 Comparison of average tensile modulus for dry and wet specimens.

respectively, which is 3.6% higher than the SBS laminate. For wet specimens, the tensile properties of BSB are 91.2 MPa and 3.57 GPa respectively which is 8.4% higher than the SBS laminate. The BSB laminate has better fibre loading and a lower percentage of resin. The absence of voids and porosity enhances the mechanical properties [12].

2.6.3 Flexural Properties of Composites

Figures 2.8 and 2.9 show the average flexural properties of dry and wet specimens. The flexural properties of SBS are 259.2 MPa and 17.5 GPa respectively, which is 44.3% higher than BSB laminate. For wet specimens the flexural properties of SBS

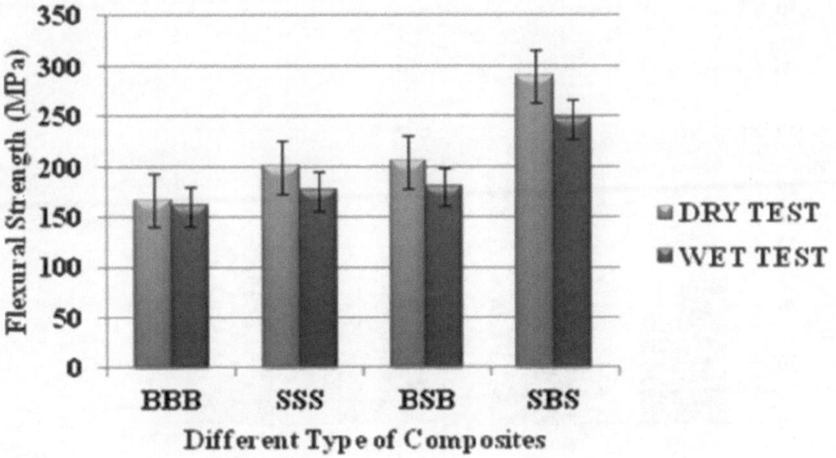

FIGURE 2.8 Comparison of average flexural strength for dry and wet specimens.

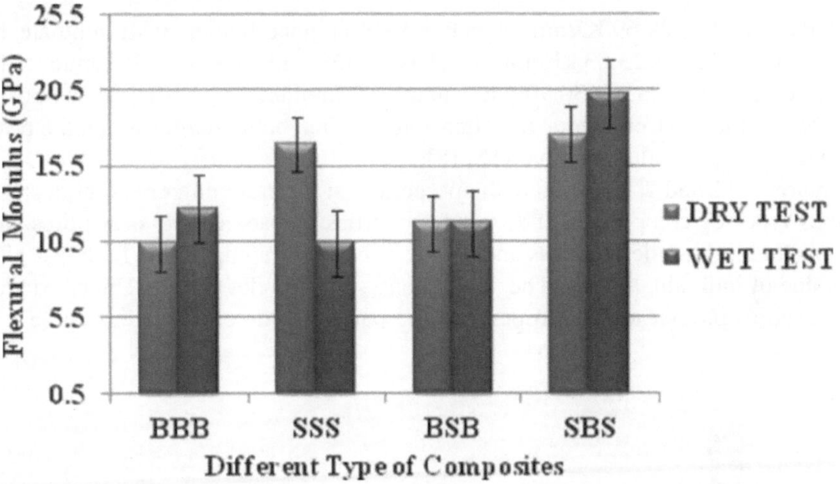

FIGURE 2.9 Comparison of average flexural modulus for dry and wet specimens.

are 247.57 MPa and 20.26 GPa respectively, which is 37.7% higher than BSB laminate. This is due to the entanglement of fibres at fibre loading and poor interfacial bonding between fibre and matrix [15, 17]. Furthermore, the sisal fibre may have undergone higher plasticity during loading, which accounts for the increased strength.

2.6.4 Impact Properties of Composites

Toughness of the material is determined by an impact test, which measures the energy absorption capability of the material when it undergoes plastic deformation. Figure 2.10 shows the impact behaviour of all types of composites. The impact

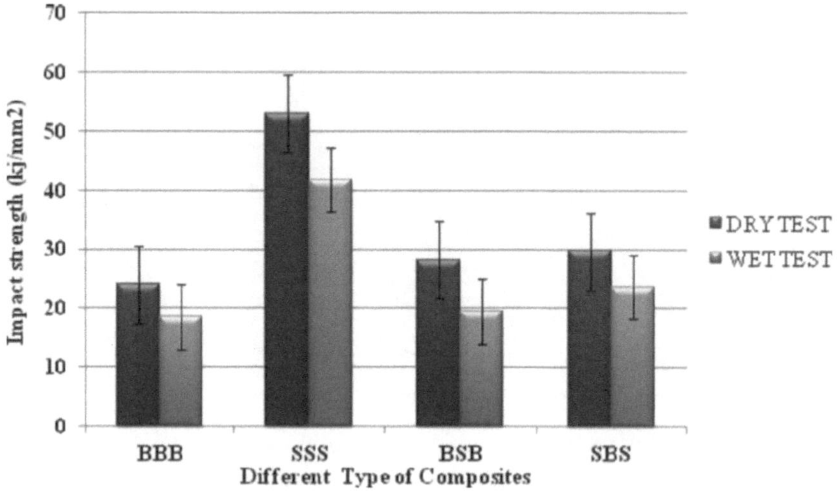

FIGURE 2.10 Comparison of impact strength for dry and wet specimens.

property of SBS is 29.69 KJ/mm^2 which is 5.36% higher than the BSB laminate. For wet types, the SBS is 23.63 kJ/mm^2 which is 21.36% higher than BSB laminate. The moisture absorption rate is 30.90% less than BSB laminate [8]. SBS type composites have better impact strength due to better fibre loading, better hardness and the brittle behaviour of lignocellulosic fibre [15, 18].

Figures 2.11 and 2.12 compare the properties of different numerical models with various types of composites. Individual properties of banana and sisal fibres were used to find the tensile modulus and tensile strength of composites. Table 2.1 gives the value of individual fibres. The parallel and series model and the Hirsch Model 1 were used to predict tensile properties. The parallel and series models agree least

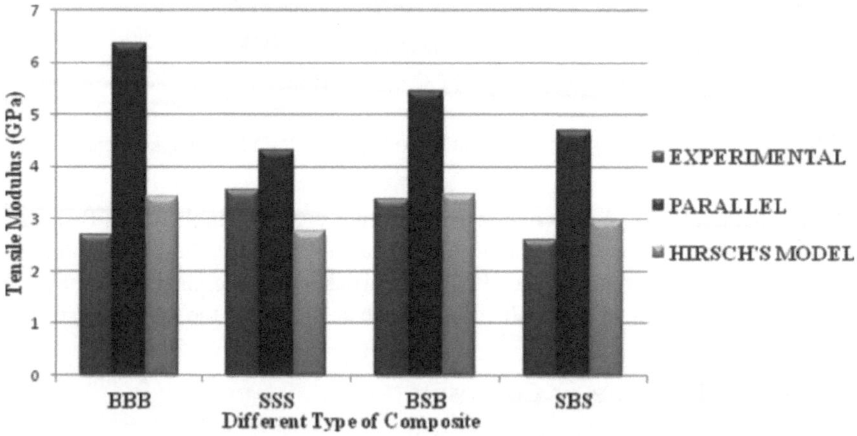

FIGURE 2.11 Tensile modulus of different type of composites.

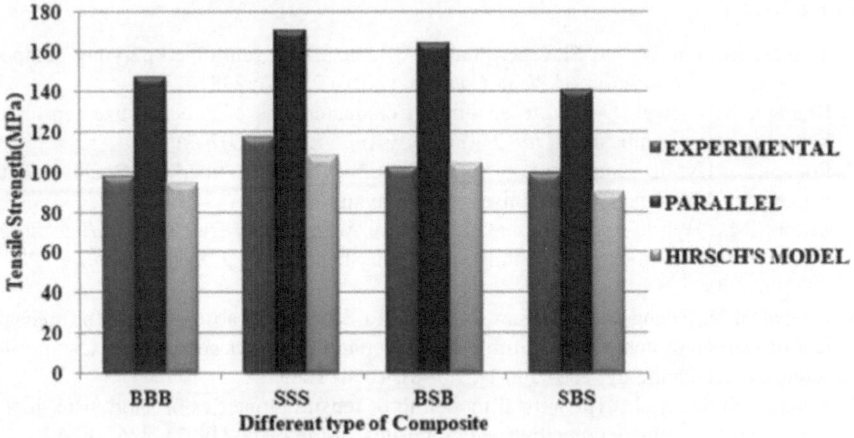

FIGURE 2.12 Tensile strength of different types of composites.

with the experimental values. Continuous fibre-reinforced polymeric composites are typically represented using parallel and series models. In reality, Hirsch's model is a hybrid of parallel and series models. Only when the value of x in Eq. (2.13 & 2.14) is 0.4 has there been agreement between theoretical and experimental values for composites that are longitudinally focused. x is a parameter that defines the stress transfer between fibre and matrix, according to this equation [9, 26, 27].

2.7 CONCLUSIONS

In this study, the mechanical properties of banana, sisal and their hybrid composites were analyzed based on the unidirectional laminate prepared (BBB, SSS, BSB, and SBS). The following conclusions can be drawn:

- Tensile strength of BSB has the highest value of 102.32 MPa compared with other types. This shows that BSB has better fibre matrix adhesion.
- Flexural strength of SBS is found to be highest at 289.86 MPa when compared with the other type of laminates. This shows that in bending mode, load transfer by sisal fibre is more efficient than other types.
- The water absorption test shows that banana fibre composite has the lowest moisture absorption percentage at 7.7%.
- Due to the inclusion of banana fibre in the composite, BBB shows the least difference when dry and wet specimens are compared.
- The impact strength of the SBS type of laminate is 34% higher than BSB laminate, due to the high impact strength of sisal fibre.
- Comparison of experimental data with theories of reinforcement shows that the value predicted by the Hirsch model aligns closely with the experimental value because it combines both parallel and series models with the introduction of factor "X".

REFERENCES

1. Venkateshwaran, N. and Elayaperumal, A. "Banana fiber reinforced polymer composites-a review", *J. Reinforced Plast. Compos.* (2010) 29(15): 2387–2396.
2. Bharath, K.N. et al. "Moisture absorption characteristics of Areca/Maize reinforced hybrid polymer composites", *Int. J. Adv. Eng. Appl.* (2010): 207–211.
3. Pothen, L. "Hybrid composites of short banana fiber and glass fiber", by Beehive digital concepts of Mahatma Gandhi University Kottayam *Polimery* (1999): 11–12.
4. Jawaid, M. "Hybrid composites made from oil palm empty fruit bunches/jute fibres: water absorption, thickness swelling and density behaviors", *J. Polym. Environ.* (2011) 19: 106–109.
5. Dhakal, H.N., Zhang, Z.A., Richardson, M.O. Effect of water absorption on the mechanical properties of hemp fibre reinforced unsaturated polyester composites. Composites science and technology. 2007 Jun 1;67(7–8): 1674–1683.
6. Kalaprasad, G. et al. "Theoretical modeling of tensile properties of short sisal fiber – reinforced low-density polyethylene composites", *Mater. Sci.* (1997): 4261–4267.
7. Sreekala, M.S., Kumaran, M.G., and Thomas, S.S. "Water sorption in oil palm fiber reinforced phenol formaldehyde composites", *Composites Part A*(2002) 33: 763–777.
8. Venkateshwaran, N., ElayaPerumal, A., Alavudeen, A., and Thiruchitrambalam, M. "Mechanical and water absorption behaviour of banana/sisal reinforced hybrid composites", *Mater. Des.* (2011) 32(7): 4017–4021.
9. Khan, T., Sultana, M.T.H., Shah, A.U.M., Ariffina, A.H., and Jawaid, M. "The effects of stacking sequence on the tensile and flexural properties of kenaf/jute fibre hybrid composites", *J. Natural Fibers* (2019): 1–12. Doi:10.1080/15440478.2019.1629148
10. Mahesha, G.T., Satish, S.B., Vijaya Kini, M., and Subrahmanya, B.K. "Mechanical characterization and water ageing behavior studies of Grewia serrulata bast fiber reinforced thermoset composites", *J. Natural Fibers* (2017): 1–13. Doi:10.1080/15440478.2017.1279103
11. Saw, S.K., Akhtar, K., Yadav, N., and Singh, A. K. "Hybrid composites made from jute/coir fibers: water absorption, thickness swelling, density, morphology, and mechanical properties", *J. Natural Fibers* (2014): 39–53. DOI:10.1080/15440478.2013.825067.
12. Dong, C., "Mechanical properties of natural fibre-reinforced hybrid composites", *J. Reinforced Plast. Compos.* (2019) 0(0): 1–13. Doi:10.1177%2F0731684419856686
13. Devireddy, S.B.R. and Biswas, S., "Physical and mechanical behavior of unidirectional banana/jute fiber reinforced epoxy based hybrid composites", *Polym. Compos.* (2017): 1396–1403. Doi:10.1002/pc.23706.
14. Ku, H., Wang, H., Pattarachaiyakoop, N., Trada, M. "A review on the tensile properties of natural fiber reinforced polymer composites", *Composites: Part B* (2011) 42: 856–873. Doi:10.1016/j.compositesb.2011.01.010.
15. ASTM Standard Tensile Test Properties D 3039.
16. ASTM Standard Flexural Test Properties D790-07.
17. ASTM Standard Impact Test ASTMD256.
18. ASTM Standard Water Absorption Test Properties D570.
19. Venkateshwaran, N., Elayaperumal, A., and Sathiya, G.K., "Prediction of tensile properties of hybrid-natural fiber composites", *Composites: Part B* 43 (2012) 793–796.
20. Paul, S.A., Boudenne, A., Ibos, L., and Candau, Y. "Effect of fiber loading and chemical treatments on thermophysical properties of banana fiber/polypropylene commingled composite materials", *Compos. A Appl. Sci. Manuf.* (2008) **39**(9): 1582–1588.
21. Joseph, S., Sreekala, M.S., Oommen, Z., and Thomas, S. "A comparison of the mechanical properties of phenol formaldehyde composites reinforced with banana fibers and glass fibers", *Compos. Sci. Technol.* (2002) 62(14): 1857–1868.

22. Dhaka, H.N., Zhang, Z.Y., and Richardson, M.O.W. "Effect of water absorption on the mechanical properties of hemp fibre reinforced unsaturated polyester composites", *Compos. Sci. Technol.* (2007) **67**(7–8): 1674–1683.
23. Melo, R.Q.C., Santos, W.R.G., Barbosa de Lima, A.G., Lima, W.M.P.B., Silva, J.V., and Farias, R.P. "Water absorption process in polymer composites: theory analysis and applications", In: Delgado J. and Barbosa de Lima A. (Eds) *Transport Phenomena in Multiphase Systems. Advanced Structured Materials*, vol. 93, Springer, Cham (2018). Doi:10.1007/978-3-319-91062-8_7.
24. Senthilrajan, S. and Venkateshwaran, N. "Ageing and its influence on vibration characteristics of jute/polyester composites", *J. Polym. Environ.* Doi:10.1007/s10924-019-01493-0.
25. Mirbagheri, J., Tajvidi, M., Hermanson, J.C., and Ghasemi, I. "Tensile properties of wood flour/kenaf fibre/polypropylene hybrid composites", *Appl. Polym. Sci.* (2007) 105(5): 3054–3059.
26. Nair, K.C.M. and Thomas, S. *J. Thermoplast. Compos. Mater.* (2003) 16: 249–271. Doi:10.1177/089270503027669
27. Fu, S.Y., Xu, G., and Mai, Y.W. "On the elastic modulus of hybrid particle/short-fiber polymer composites", *Compos B – Eng.* (2002) 33(4): 91–299
28. Dhakal, H.N. et al. "Effect of water absorption on the mechanical properties of hemp fiber reinforced unsaturated polyester composites", Advanced Polymer and Composites (APC) Research Group, 12 April 2006; received in revised form 22 June 2006; accepted 29 June 2006.

3 Interfacial Adhesion Improvement of Polymer Composites Using Graphene Fillers

Furkan Ahmad
Delhi Technological University, Shahbad Daulatpur, New Delhi, India

Vijay Chaudhary
Amity University, Noida, Uttar Pradesh, India

Ankit Manral
Netaji Subhas University of Technology (Formerly NSIT), Dwarka, New Delhi, India

CONTENTS

3.1 INTRODUCTION

Although much work has been done to improve the mechanical properties of polymer nanocomposites, these properties remain modest compared to those of conventional materials. Young's modulus of amorphous polymer composites has been found to be single-digit GPa in completely cured state [1]. Moreover, the strength of these polymer composites decreases non-linearly when they are used at higher temperature

DOI: 10.1201/9781003128861-3

around the glass transition temperature, which is usually close to the ambient temperature [2]. It has therefore become necessary to introduce some kind of organic/inorganic filler to improve the structural stability of the developed composites. Glass, metal and semiconducting fillers are used to critically control the structural strength along with other required properties such as light weight, chemical inertness and multifunctionality of the material. Properties of these advanced nanocomposites, including structural strength, can be tailored according to various applications.

Indeed, the structural strength of the nanocomposite is dependent on the interfacial area and the strength of the interface. Nanocomposites with nanofillers are a novel class of materials with increased interfacial strength due to their increased interfacial area and controllable interactions at the interface. A critical factor for polymeric composite materials is the difference in physical components present in the composite materials. Nanofillers are agents that are used to reduce the risk of physical mismatch between nanofiller and matrix material by increasing the interfacial area at the nano scale. Nanoscale fillers are used in the polymer composite materials field to improve every aspect of performance: structural stability, strength, or multifunctionality such as thermal or electrical conductivity or improved glass transition temperature [2]. Nanocomposites not only improve structural strength but also greatly improve functional properties.

Graphene and its derivatives have shown enormous potential for innovatively improving the mechanical and structural strength of reinforced polymer composites. In the present chapter, if not otherwise specified, graphene means not only traditional graphene materials but also derivatives of graphene. Graphene–polymeric composite materials are not used for their improved mechanical and structural strength but also other functional properties, such as optical transportation [3–6]. They are used in a wide range of industrial, domestic and sophisticated engineering and medical applications.

The number of publications related to graphene–polymer nanocomposites has grown exponentially in the past decade (Figure 3.1), although no significant records were found before 2010. Most of the fundamental properties of graphene and its derivatives have been explored and the use of graphene–polymer composite materials has increased manyfold in the past decade. While the use of graphene materials as a nanofiller has proved excellent for polymer and other composite materials, current research is focusing on the development of totally new materials using only graphene and its derivatives for atomic-level multi-stacking of 2D layers [2].

In nanocomposite materials, filler material is supposed to dominate all the properties but, in fact, interfacial strength affects the properties to a great extent. Dispersion of the nanofiller is an aggressive method of controlling the interfacial area. Poor interfacial strength leads to lower mechanical and structural stability of components developed using nanomaterial fillers because of the extensive aggregation and inefficient dispersion of the filler materials. These problems are always associated with graphene–polymer composite materials. Elastic modulus, toughness and tensile strength are the main parameters by which performance of nanocomposite materials can be judged [7]. Most applications require a balance between all these properties. Toughness of the material is normally evaluated by the stress-strain area. Stiffened

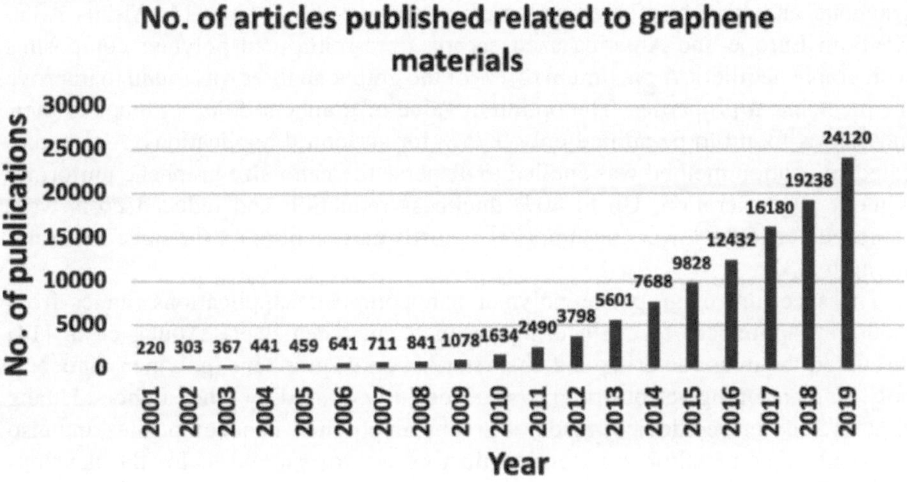

FIGURE 3.1 Research trend in the field of graphene-related materials and applications.

nanofillers can alter the toughness of resultant material with strong filler–matrix interactions but can also reduce the elongation. However, overdose of filler materials can lead to slippage at the filler–matrix interface leading to reduced toughness [2]. Graphene derivatives, on the other hand, provide a perfect filler as they are both flexible and strong enough for structural applications [8, 9]. Among all the graphene derivatives, graphene oxide fillers are compatible with most polymers, proving to be a good filler irrespective of the matrix material. Thin sheets of graphene oxide can be assembled into "paper" form using various techniques [10, 11]. Elastic modulus of around 40 GPa and toughness of 0.26 MJ m^{-3} can be obtained using these "paper" materials. However, these values are lower than the predicted mechanical values of graphene oxide materials [8]. Borate-assisted crosslinking has been proved to provide much stronger graphene oxide paper but the brittleness of the resulting materials also increases, reducing its application spectrum. Some authors have suggested using additional crosslinking of the graphene oxide layers in the multi-layer graphene oxide thin sheets [10, 12]. Some authors have used esterification as a successful method to improve crosslinking of graphene oxide flakes [10]. A number of other authors have reported improvement in the structural strength of graphene nanocomposite materials that can be used in domestic and industrial applications.

3.2 GRAPHENE: APPLICABILITY SPECTRUM

The journey of graphene from laboratory to industrial application has been very swift and smooth, thanks to its remarkable mechanical and other overall properties. In 2018, Professor Ahmed Elmarakbi of the Department of Computing, Engineering and Technology at the University of Sunderland, UK showcased the world's first graphene composite structural component for the automobile industry. Professor Elmarakbi fabricated a car bumper with improved mechanical properties using graphene and fibre-reinforced polymer composites [13]. In 2014, Airbus A350 XWB in Europe and America used carbon fibre-reinforced polymer composites with graphene fillers. A maximum of 1% of the graphene filler was found to improve the mechanical properties. The optimum value of graphene filler with epoxy thermoset was found to be around only 0.75% for structural applications. A new patented sonication method was applied to disperse the nano-size graphene uniformly without agglomeration. Up to 40% thickness reduction and reduced costs were achieved due to the lower consumption of costly carbon fibre for the same structural strength [13].

The spectrum of graphene–polymer nanocomposite applications ranges from industrial appliances to engineering products. In recent work, Young et al. [14] discussed the manufacturing and characterization of graphene–polymer nanocomposites, mentioning various products successfully crafted by other authors. Huang et al. [15] described devices made of graphene–polymer nanocomposites and also explored other metal/organic/semiconducting constituents used for the development of various industrial devices. Recently, Sun et al. [16] explored various important parameters affecting the integration of graphene and carbon nanotubes for the manufacturing of polymer nanocomposites. Yang et al. [17] fabricated thin sheets of nanocomposites using a layer-to-layer stacking method for industrial applications. Lee et al. [18] measured the modulus of elasticity of a monolayer graphene material sheet and found it to be highest at 1TPa, while Balandin et al. [19] achieved very high thermal conductivity of 5.1 ×_103 W m^{-1} K^{-1} for single-layer graphene material. Du et al. [20] approached a ballistic electricity transport at the rate of 6 × _105 S m^{-1} in the suspended graphene materials. Graphene is also regarded as the thinnest 2D nano filler, a single sheet having a thickness of 0.34 nm with high intrinsic flexibility [21].

These exceptional properties make graphene quite popular among materials scientists. However, various derivatives of graphene that possess partial excellent properties of graphene are attracting more scientific attention than pure graphene. One such popular graphene derivative is graphene oxide which can provide excellent mechanical and chemical properties. An important characteristic of graphene oxide is biocompatibility [22, 23]. Abundant oxygen present in the graphene oxide makes it suitable for transformation into other groups like carboxyl, hydroxyl and carbonyl [24]. As we know, covalent bonding is the strongest of the various types of bonding present at molecular level. Covalent grafting of graphene oxide can help to achieve good mechanical strength by improving the blending of oxide components and matrix components [10]. The electrostatic interactions present in the nanocomposites with graphene oxide fillers, together with the restorability of these interactions, make them stronger than other nanocomposites without graphene oxide fillers [25].

3.2.1 Graphene: Improved Mechanical Strength

Modulus of elasticity, tensile strength, flexural strength and thermal stability are properties required for any structural application. Much research has been published on improving the properties of graphene and its derivative-based nanocomposites. Polyallylamine (PAA) has been proved to significantly improve the elastic modulus of nanocomposites with graphene [11]. Extensive sonication was required to obtain a homogeneous mixture of graphene oxide and PAA. The highest modulus of elasticity obtained by using PAA modifiers was found to be 33 GPa, which is higher than in the case of nanocomposites without PAA modifiers. The reinforcing effect of PAA was found to be modest with respect to other effects reported in other literatures. Tian et al. [12] used 30% polyethyleneimine (PEI) in place of PAA to crosslink graphene oxide materials. The developed materials showed elastic modulus of around 100 GPa and exceptional mechanical strength of 210 MPa. However, these nanocomposites reduced the ultimate strain by 0.2%. Toughening of the ultra-thin graphene oxide membrane layer was obtained by electrostatic interactions at the interface. A huge jump of 500% in the toughness of the membrane from 0.4 MJ m^{-3} to 1.9 MJ m^{-3} was recorded [25, 26]. The interface area of the two constituents increased significantly, resulting in better transfer of the stress from one constituent to the other in loading conditions. The optimum value of the graphene oxide for the high toughness value was found to be only 3.3% by volume. Consequently, the Young's modulus of the developed material was increased eight times up to a value of 18 GPa. The ultimate stress and ultimate strain value also increased by 120% and 50% respectively. The increment in the value of the ultimate strain was quite surprising as the graphene oxide generally led to increased brittleness of the materials, reducing the ultimate strain. In this particular case, the interfacial interactions were stated to be either too strong or too weak, facilitating stress distribution.

Graphene oxide behaves differently with different matrix materials. With the polyvinyl alcohol (PVA) matrix, the graphene oxide makes a strong hydrogen bond at the interface, which controls the mechanical properties of the developed nanocomposites. On the other hand, the hydrogen bond formed at the interface of the PMMA matrix and graphene oxide is weaker, resulting in modest mechanical properties of Young's modulus of 6 GPa and ultimate strain increased up to a value of 2.6% [2]. In another research study, Li et al. [27] added only 3% graphene oxide to the PVA matrix and concluded that the storage modulus of the material was moderately improved by 50%, achieving a value of 6 GPa, but the ultimate strength was increased by 100%, giving a value of 60 GPa with minimal reduction in the ultimate strain. In most of the available literature, the mechanical performance of these graphene nanocomposites was higher than the performance of the individual components, especially the Young's modulus which was found to have higher values than predictions based on analytical and simulated models. This improvement in mechanical performance was considered to be due to heterogeneous interactions at the interface of the matrix and graphene oxide.

3.2.2 Graphene: Improved Thermal Stability

Thermal stability of the structural components is also an important factor to consider in design and selection of materials. Graphene improves thermal stability of the polyvinyl chloride (PVC) matrix to a great extent by increasing the glass transition temperature of thin sheets fabricated with graphene flakes. These ultra-thin nanocomposite sheets were also tested for mechanical, electrical and thermal performance. With a very low loading of graphene flakes, the ultimate tensile strength and the modulus of elasticity was found to improve. Storage modulus increased by five times in the dynamic mechanical analysis of the developed PVC sheets. Graphene fillers were prepared by exfoliation assisted by CTAB. Electrical as well as mechanical properties were found to be improved when graphene flakes were used as nanofillers [28]. The robustness of the graphene and its derivatives has increased demand, while numerous environmental issues have been addressed. Yousuf et al. [29] reviewed the available literature on the applications of graphene and graphene-based nanocomposites. Among environmental applications, graphene and its derivatives are being used as absorbents and detectors. Detection of pollutants (organic and inorganic) from waste water has become possible due to innovations in the field of graphene nanocomposites. These materials are also constantly used in various energy storage appliances such as batteries, fuel cells and capacitors, due to the extraordinary chemical structure of graphene. Graphene and its derivatives are also used for the production and storage of hydrogen.

Lightweight composite materials are a vital innovation that can improve fuel efficiency of vehicles up to 8% for each 10% reduction in weight, making the automobile sector more competitive [30]. Graphene and its derivatives have opened a new and promising window for the development of lightweight materials for automobile structures.

Graphene with carbon-fibre polymer composites is an active area of research for the development of structural components. These composites are generally known as hybrid composites. Possible applications of these hybrid composites include wind turbine blades, lightweight beams and super-light cars. The large surface area available for bonding makes these nano materials inevitable in applications like lightweight beams. Carbon fibres at the macro scale along with graphene or nanofibres at the micro scale can be used for structural components [31].

Graphene has proved an excellent nanoscale material with a wide range of structural applications in the scientific community, including tissue engineering. While graphene oxide-based biosensors have been used for quick diagnosis of numerous diseases, biocompatibility of the graphene and its derivatives is essential for positive results in the field of human health [32]. A lot of research has been done to improve the structural strength of graphene-based nanocomposites. Srivastava et al. [33] coated carbon fibre with graphene nanoplatelets (GNPs) and found the flexural strength of carbon-fibre composites was improved with a continuous interphase formed between fibre and matrix. Oxygen functional groups enhanced the bonding between carbon fibre and matrix, leading to a dramatic improvement in flexural strength. High surface roughness increased mechanical interlocking at the interphase, further improving the flexural strain. Cha et al. [34] reinforced the epoxy matrix with melamine functionalized carbon nanotubes (M-CNTs) and graphene nanoplatelets (M-GNPs). Functionalized graphene with epoxy showed exceptional performance in terms of elastic modulus, ultimate tensile strength and fracture toughness which increased by 71%, 23% and 124% respectively. 2D nanoplatelets offered larger surface area and better interfacial strength than 1D nanotubes, which helped in bridging the crack and led to better transfer of load from matrix to reinforcement.

3.2.3 Graphene: Next-Gen Materials

Lightweight and next-generation materials have become possible with advancements in the field of graphene and the functionalization of graphene derivatives. Modification of the matrix materials with reduced graphene oxide (rGO) is quite challenging as the dispersion of the nanoscale-reduced graphene oxide is not easy. The content of rGO in epoxy increases the viscosity of the matrix which further restricts the flow during processing of the nanocomposites. The interfacial adhesion between reinforcement and matrix improved with the addition of reduced graphene oxide in the carbon fibre-reinforced epoxy composites. Incorporation of the rGO content increased interlaminar shear strength, fracture toughness and impact strength of the developed composites [35]. Kim et al. [36] demonstrated excellent enrichment in the mechanical properties of reduced graphene oxide (rGO) fibres after PDA infiltration and pyrolysis. The PDA was transformed into N-doped graphene layers to fabricate innovative py-PDA-rGO fibres. These fibres provide adhesive properties for enriched mechanical strength, while at the same time connecting the rGO flakes for improved electrical properties. Mechanical strength was improved 3.56 times while elastic modulus was improved 3.95 times. Electrical conductivity was increased almost 10-fold by the new innovative fibre reinforcement. In a study, Yao et al. [37] demonstrate an appealing method for dispersing nanosheets of graphene in epoxy matrix using chemical functionalization with 4-nitrobenzenediazonium salt. The above method proved to be good for homogeneous mixing of graphene nanosheets with high stability, improving the mechanical and thermal properties of the parent matrix epoxy. Only 0.8 wt.% of the graphene nanosheets improved elongation at break and tensile strength of epoxy-based nanocomposites by 63% and 37% respectively compared to neat epoxy. Thermal conductivity of neat epoxy was also improved 2.5 times with only 5 wt.% loading of the dispersed graphene. Mechanical properties were enhanced with efficient use of carbon nanotubes (CNTs), using functionalization of graphene oxide along with a quaternary ammonium group designed QA-FGO. Other researchers [38] claim that the strongest interactions of π-cation were induced in the fibres to improve tensile strength, and modulus and toughness from 58 MPa, 2.5 GPa and 1.1 MJ/m^3 to 409 MPa, 13.4 GPa and 10.1 MJ/m^3 respectively. Incorporation of the graphene nanoplatelets into the epoxy matrix encouraged the improvement of fatigue resistance by introducing multiple toughening mechanisms. Crack deflection, debonding of the nanofillers, crack bridging using nanofillers and crack tip shielding are the main mechanisms behind the improved fatigue resistance. Agglomerated nanofillers might hinder the toughening of the materials while alignment of the GNPs in the direction of the crack may effectively toughen the matrix materials [39].

Non-uniform but controlled dispersion of the graphene-based nanofiller to make functionally graded composite materials has opened up a new pathway for development of next-generation materials. The dispersion occurs in such a way that the quantity of the nanofillers is reinforced in the area where it is most required, depending on the loading conditions and stresses developed in the component. An excellent combination of multifunctionality, high stiffness and super light weight can be achieved in these next-generation functionally graded materials. Key challenges in the field of functionally graded materials are the optimization of the design and an efficient processing technique for successful fabrication of the structural components. The number of publications in the field of graphene-reinforced functionally graded materials has increased exponentially in the last three years, with China and Iran the top contributors [40]. In order to achieve better compatibility of oxidized expanded graphene nanoplatelets (EGNPs) with epoxy matrix, functionalization of EGNPs with amine was successfully completed. Raman spectroscopy and Fourier Transform Infrared spectroscopy were used to verify the presence of functional groups. Increments of up to 66% were recorded in the fracture toughness value of nanocomposites, compared to neat epoxy with only 0.1 wt.% loading of the EGNPs. Difficulties in the homogeneous dispersion of the nanoscale EGNPs was the main problem encountered in processing these nanocomposites. The use of fully exfoliated graphene 2D sheets has been proposed for the production of ultra-strong materials for high-end engineering structural applications [41].

3.2.4 Graphene: Improved Fracture Strength

Ahmadi-Moghadam and Taheri [42] fabricated GNP of average size 7 nm and epoxy-based nanocomposites, using a mechanical stirrer rotating at 2000 rpm for 15 min of operation time to properly mix the GNP nanoparticles in the epoxy matrix. The mode I and mode II fracture toughness of developed nanocomposites were evaluated. Mode I fracture toughness varied linearly with the weight fraction of the graphene nanoplatelets, while mode II fracture toughness reduced on loading the graphene nanoplatelets [42]. These authors concluded that loading of graphene nanofillers increased plastic deformation, creating the shear bands for mode I failure. Surface roughness also improved with the increased loading of nanocomposites. Crack deflection was the main factor to be considered for the toughening of these nanocomposite materials. Fracture toughness of the epoxy was improved by 41% by adding just 0.5% graphene by weight. No increase was recorded after 0.75% addition of graphene by weight [43]. Thermal conductivity of the epoxy matrix was increased by 306% by loading only 10 wt.% of graphene nanoplatelets (GNPs). Although higher GNP content seems to improve the thermal conductivity of the epoxy matrix, using a limited amount of GNPs due to their negative effect on the contact angle was proposed [44]. The thermal stability of the neat epoxy was improved to a great extent by using only 0.25% of the GNS loading by weight [45]. Intercalated graphene influenced the properties of nanocomposites more than graphene flakes [46]. Improved glass transition temperature was observed in the intercalated graphene-reinforced polymer composites. Various industrial applications, such as solar cells and supercapacitors, are being fabricated using graphene nanocomposites [47]. Graphene quantum dots (GQDs) have been fabricated for applications such as LEDs, bioimaging and photovoltaic devices. The on/off current ratio of the TFT fabricated using GQDs-ZnO was improved significantly in comparison to TFTs with pure ZnO [48]. Many applications are being realized based on improved performances of polymer composites by introducing graphene and its derivatives (Table 3.1).

TABLE 3.1
Structural and Some other Applications of Graphene-based Polymer Composites

Sr. No.	Graphene constituents	Remarks	Applications	Reference
1.	Graphene oxide (GO), Reduced graphene oxide (rGO)	Only 3% loading of rGO showed 47% reduced oxygen permeability	Structural sheets for gas transportation	[49, 50]
2.	Graphene	Structural component for automobile industry	A car front bumper	[13]
3.	Reduced graphene oxide (rGO)	Energy storage application for electronics	Supercapacitors, lithium ion batteries, stretchable electronics	[51, 52]
4.	Graphene platelets	Up to 52% increase in buckling load with only 0.1% addition of graphene platelets	Lightweight buckling-resistant structural elements for aerospace	[53]
5.	Graphene/ Graphene oxide	Exceptional structural and electronic properties	Highly sensitive biosensors	[54]
6.	Functionalized graphene	Nanocomposites for environmental applications	Gas sensors for organic and inorganic vapours, bacterial detection and removal	[55]
7.	Graphene	Structural components for wind energy applications	Wind turbine blade	[56]

3.3 CHALLENGES

Despite the improvements in the mechanical properties of graphene-based nanocomposites resulting from a combination of various technologies, there are still a number of challenges to be considered before they can be used in practice in industrial applications. These include:

1. Large-scale production of graphene and its derivatives is quite complex as exfoliation methods cannot be scaled to industry level.
2. Improving the present method of functionalization of graphene and its derivatives remains a challenge.
3. Reduced ductility of graphene-reinforced nanocomposite materials also poses difficulties while synthesizing materials for structural applications, especially for the automotive and aerospace industries.
4. Although a lot of research is being done on characterization of graphene-reinforced polymer nanocomposites, the available literature provides little insight into the various factors affecting its strength.

5. Existing material models for the modelling of graphene-based polymer nanocomposites are inadequate and cannot efficiently predict properties for high-end structural applications.
6. Insufficient knowledge of the processing of automobile components which can produce high-performing structures.
7. Problems in the joining of graphene-reinforced polymer composites are yet to be solved by researchers.
8. Life cycle assessment (LCA) of the graphene and its derivative-based products should be done to evaluate the environmental impact.

Graphene and its derivatives have been proved to be outstanding nanofillers providing unmatched mechanical and electrical properties for structural applications. Nevertheless, true industrial applications have yet to become a reality due to insufficient knowledge of the proper dispersion of the nanofiller and control of nanoscale phenomena such as ordering of graphene chains, stretching and orientation.

3.4 CONCLUSIONS

This chapter has summarized recent efforts in the field of structural applications of graphene and its derivatives-based polymer nanocomposites. The structural stability and mechanical strength of polymer matrix-based reinforced composites can be improved to a great extent by using graphene while simultaneously reducing component weight. The available processing methods are not efficient for mass-scale production of graphene and its derivatives-based nanocomposites. Proper dispersion of nanoscale graphene is a necessary condition for the production of homogeneous properties. Currently, a number of applications for high-performing graphene–polymer nanocomposites are not being realized. Finally, the unavailability of numerical models for finite element analysis of structural components is a vital hurdle impeding industrialization of graphene-based products. Model assumptions currently available are vague and are unable to satisfactorily predict the properties of developed nanocomposite materials.

REFERENCES

1. Sperling LH. *Introduction to Physical Polymer Science*. 4th edition, Hoboken, NJ: John Wiley & Sons Inc.; 2006, 845.
2. Hu K, Kulkarni DD, Choi I, Tsukruk VV. Graphene-polymer nanocomposites for structural and functional applications. *Prog Polym Sci* 2014; 39: 1934–1972.
3. Britnell L, Ribeiro RM, Eckmann A, Jalil R, Belle BD, Mishchenko A, Kim YJ, Gorbachhev RV, Georgiou T, Morozov SV, Grigorenko AN, Geim AK, Casiraghi C, Castro Meto AH, Novoselov KS. Strong light-matter interactions in heterostructures of atomically thin films. *Science* 2013; 340: 1311–1314.
4. El-Kady MF, Kaner RB. Scalable fabrication of high-power graphene micro-supercapacitors for flexible and on-chip energy storage. *Nat Commun* 2013; 4: 1475–1481.
5. Kim KS, Zhao Y, Jang H, Lee SY, Kim JM, Kim KS, Ahn JH, Kim P, Choi JY, Hong BH. Large-scale pattern growth of graphene films for stretchable transparent electrodes. *Nature* 2009; 457: 706–710.

6. Tetsuka H, Asahi R, Nagoya A, Okamoto K, Tajima I, Ohta R, Okamoto A. Optically tunable amino-functionalized graphene quantum dots. *Adv Mater* 2012; 24: 5333–5338.
7. Fischer H. Polymer nanocomposites: from fundamental research to specific applications. *Mater Sci Eng C* 2003; 23: 763–772.
8. Stankovich S, Dikin DA, Dommett GHB, Kohlhaas KM, Zimney EJ, Stach EA, Piner RD, Nguyen ST, Ruoff RS. Graphene-based composite materials. *Nature* 2006; 442: 282–286.
9. Mamedov AA, Kotov NA. Free-standing layer-by-layer assembled films of magnetite nanoparticles. *Langmuir* 2000; 16: 5530–5533.
10. Cheng Q, Wu M, Li M, Jiang L, Tang Z. Ultratough artificial nacre based on conjugated cross-linked graphene oxide. *Angew Chem Int Ed* 2013; 52: 3750–3755.
11. Park S, Dikin DA, Nguyen ST, Ruoff RS. Graphene oxide sheets chemically cross-linked by polyallylamine. *J Phys Chem C* 2009; 113: 15801–15804.
12. Tian Y, Cao Y, Wang Y, Yang W, Feng J. Realizing ultrahigh mod-ulus and high strength of macroscopic graphene oxide papers through crosslinking of mussel-inspired polymers. *Adv Mater* 2013; 25: 2980–2983.
13. Mathijsen D. Graphene flagship yields breakthrough in composites technology for structural applications. *Reinf Plast* 2018; 62: 132–137.
14. Young RJ, Kinloch IA, Gong L, Novoselov KS. The mechanics of graphene nanocomposites: a review. *Compos Sci Technol* 2012; 72: 1459–1476.
15. Huang X, Yin Z, Wu S, Qi X, He Q, Zhang Q, Yan Q, Boey F, Zhang H. Graphene-based materials: synthesis, characterization, properties, and applications. *Small* 2011; 7: 1876–1902.
16. Sun X, Sun H, Li H, Peng H. Developing polymer composite materials: carbon nanotubes or graphene? *Adv Mater* 2013; 25: 5153–5176.
17. Yang M, Hou Y, Kotov NA. Graphene-based multilayers: critical evaluation of materials assembly techniques. *Nano Today* 2012; 7: 430–447.
18. Lee C, Wei X, Kysar JW, Hone J. Measurement of the elastic properties and intrinsic strength of monolayer graphene. *Science* 2008; 321: 385–388.
19. Balandin AA, Ghosh S, Bao W, Calizo I, Teweldebrhan D, Miao F, Lau CN. Superior thermal conductivity of single-layer graphene. *Nano Lett* 2008; 8: 902–907.
20. Du X, Skachko I, Barker A, Andrei EY. Approaching ballistic transport in suspended graphene. *Nat Nanotechnol* 2008; 3: 491–495.
21. Allen MJ, Tung VC, Kaner RB. Honeycomb carbon: a review of graphene. *Chem Rev* 2009; 110: 132–145.
22. Liu Y, Yu D, Zeng C, Miao Z, Dai L. Biocompatible graphene oxide-based glucose biosensors. *Langmuir* 2010; 26: 6158–6160.
23. Wang K, Ruan J, Song H, Zhang J, Wo Y, Guo S, Cui D. Biocompatibility of graphene oxide. *Nanoscale Res Lett* 2011; 6: 1–8.
24. Pei S, Cheng HM. The reduction of graphene oxide. *Carbon* 2012; 50: 3210–3228.
25. Kulkarni DD, Choi I, Singamaneni SS, Tsukruk VV. Graphene oxide-polyelectrolyte nanomembranes. *ACS Nano* 2010; 4: 4667–4676.
26. Ramanathan T, Abdala AA, Stankovich S, Dikin DA, Alonso MH, Piner RD, Adamson DH, Schniepp HC, Chen X, Ruoff RS, Nguyen ST, Aksay IA, Prud'Homme RK, Brinson LC. Functionalized graphene sheetsfor polymer nanocomposites. *Nat Nano Technol* 2008; 3: 327–331.
27. Li Z, Young RJ, Kinloch IA. Interfacial stress transfer in graphene oxide nanocomposites. *ACS Appl Mater Interfaces* 2013; 5: 456–463.
28. Vadukumpully S, Paul J, Mahanta N, Valiyaveettil S. Flexible conductive graphene/poly(vinyl chloride) composite thin films with high mechanical strength and thermal stability. *Carbon* 2011; 49: 198–205.

29. Yusuf M, Kumar M, Khan MA, Sillanpaa M, Arafat H. A review on exfoliation, characterization, environmental and energy applications of graphene and graphene-based composites. *Adv Colloid Interface Sci* 2019; 273: 1–23. doi:10.1016/j.cis.2019.102036.
30. Joost WJ. Reducing vehicle weight and improving US energy efficiency using integrated computational materials engineering, *JOM* 2012; 64: 1032–1038, doi:10.1007/s11837-012-0424-z.
31. Hallad SA, Banapurmath NR, Hunashyal AM, Shettar AS, Ayachit NH, Mruthunjaya AK, Lohit RB, Uttur M. Experimental investigation for graphene and carbon fibre in polymer-based matrix for structural applications. *J Appl Res Technol* 2017; 15: 297–302.
32. Prabhakar O, Arun K. Recent progresses and challenges in graphene based nano materials for advanced therapeutical applications: a comprehensive review. *Mater Today Commun* 2020; 22: 100823.
33. Srivastava AK, Gupta V, Yerramalli CS, Singh A. Flexural strength enhancement in carbon-fiber epoxy composites through graphene nano-platelets coating on fibers. *Compos Part B* 2019; 179: 107539.
34. Cha J, Kim J, Ryu S, Hong SH. Comparison to mechanical properties of epoxy nanocomposites reinforced by functionalized carbon nanotubes and graphene nanoplatelets. *Compos Part B* 2019; 162: 283–288.
35. Adak NC, Chhetri S, Kuila T, Murmu NC, Samanta P, Lee JH. Effects of hydrazine reduced graphene oxide on the inter-laminar fracture toughness of woven carbon fiber/epoxy composite. *Compos Part B* 2018; 149: 22–30.
36. Kim J, Hwang H, Yoo SC, Seo H, Ryu S, Hong SH. Effect of pyrolyzed catecholamine polymers for concurrent enhancements of electrical conductivity and mechanical strength of graphene-based fibers. *Compos Sci Technol* 2018; 183: 107818.
37. Yao H, Hawkins SA, Sue H-J. Preparation of epoxy nanocomposites containing well-dispersed graphene nanosheets. *Compos Sci Technol* 2017; 146: 161–168.
38. Kim Y-J, Park J, Yang C-M, Jeong HS, Kim SM, Han SW, Yang B, Kim Y-K. Bio-inspired incorporation of functionalized graphene oxide into carbon nanotube fibers for their efficient mechanical reinforcement. *Compos Sci Technol* 2019; 181: 107680.
39. Bhasin M, Wu S, Ladani RB, Kinloch AJ, Wang CH, Mouriţz AP. Increasing the fatigue resistance of epoxy nanocomposites by aligning graphene nanoplatelets. *Int J Fatigue* 2018; 113: 88–97.
40. Zhao S, Zhao Z, Yang Z, Ke LL, Kitipornchai S, Yang J. Functionally graded graphene reinforced composite structures: a review. *Eng Struct* 2020; 210: 110339.
41. Chatterjee S, Wang JW, Kuo WS, Tai NH, Salzmann C, Li WL, Hollertz R, Nuesch FA, Chu BTT. Mechanical reinforcement and thermal conductivity in expanded graphene nanoplatelets reinforced epoxy composites. *Chem Phys Lett* 2012; 531: 6–10.
42. Ahmadi-Moghadam B, Taheri F. Fracture and toughening mechanisms of GNP-based nanocomposites in modes I and II fracture. *Eng Fract Mech* 2014; 131: 329–339.
43. Cha J, Kim J, Ryu S, Hong SH. Comparison to mechanical properties of epoxy nanocomposites reinforced by functionalized carbon nanotubes and graphene nanoplatelets. *Compos Part B* 2019; 162: 283–288.
44. Moriche R, Prolongo SG, Sánchez M, Jiménez-Suárez A, Chamizo FJ, Urena A. Thermal conductivity and lap shear strength of GNP/epoxy nanocomposites adhesives. *Int J Adhes Adhes* 2016; 68: 407–410.
45. Chhetri S, Adak NC, Samanta P, Murmu NC, Hui D, Kuila T, Lee JH. Investigation of the mechanical and thermal properties of L-glutathione modified graphene/epoxy composites. *Compos Part B* 2018; 143: 105–112.

46. Shiu S-C, Tsai J-L. Characterizing thermal and mechanical properties of graphene/epoxy nanocomposites. *Compos Part B* 2014; 56: 691–697.
47. Lawal AT. Graphene-based nano composites and their applications. A review. *Biosens Bioelectron* 2019; 141: 111384.
48. Zhu H, Liu A, Xu Y, Shan F, Li A, Wang J, Yang W, Barrow C, Liu J. Graphene quantum dots directly generated from graphite via magnetron sputtering and the application in thin-film transistors. *Carbon* 2015; 88: 225–232.
49. Yang Y-H, Bolling L, Priolo MA, Grunlan JC. Super gas barrier and selectivity of graphene oxide-polymer multilayer thin films. *Adv Mater* 2013; 25: 503–508.
50. Kausar A. Composite coatings of polyamide/graphene: microstructure, mechanical, thermal, and barrier properties, *Compos Interfac* 2018; 25: 109–125.
51. Chee WK, Lim HN, Harrison I, Chong KF, Zainal Z, Ng CH, Huang NM. Performance of flexible and binderless polypyrrole/graphene oxide/zinc oxide supercapacitor electrode in a symmetrical two-electrode configuration, *Electrochim Acta* 2015; 157: 88–94.
52. Lin D, Liu Y, Liang Z, Lee H-W, Sun J, Wang H, Yan K, Xie J, Cui Y. Layered reduced graphene oxide with nanoscale interlayer gaps as a stable host for lithium metal anodes, *Nat Nanotechnol* 2016; 11: 626–632.
53. Rafiee MA, Rafiee J, Yu ZZ, Koratkar N. Buckling resistant graphene nanocomposites. *Appl Phys Lett* 2009; 95: 223103.
54. Zhang M, Li Y, Su Z, Wei G. Recent advances in the synthesis and applications of graphene–polymer nanocomposites. *Polym Chem* 2015; 6: 6107–6124.
55. Chang H, Wu H. Graphene-based nanocomposites: preparation, functionalization, and energy and environmental applications. *Energy Environ Sci* 2013; 6: 3483–3507.
56. Kumar A, Sharma K, Dixit AR. A review of the mechanical and thermal properties of graphene and its hybrid polymer nanocomposites for structural applications. *J Mater Sci* 2019; 54: 5992–6026.

4 Failure Models of Composite Structures under Impact Loading

Priyanka and N. Shanmuga Priya
Siddaganga Institute of Technology, Tumkuru

M. Shanmugasundaram
Jazan University, Kingdom of Saudi Arabia

CONTENTS

4.1 INTRODUCTION

Glass/polyester laminated structural materials are widely used in the aviation, shipbuilding and land transport industries, due to their suitable mechanical characteristics, transparency to electromagnetic waves and low product cost. Although these laminated structures are not intended to be armour, they may suffer impacts. Impact may affect various properties of laminated composites (Zhou 1995). It can significantly diminish their strength and this may not be detected visually (Ibekwe et al. 2007), which is the main reason limiting the use of laminate-type composite materials (Hawyes et al. 2001). This chapter discusses damage and impact load on composite materials, the classification of impacts, impact response, impact force on composites and their failure modes.

DOI: 10.1201/9781003128861-4

4.2 DAMAGES IN COMPOSITE MATERIAL

In-service conditions can subject composites to unexpected damage due to impact load, excess heat, ultra-violet light and runway bird strikes. Heat and ultraviolet can degrade composites by initiating oxidation, which reduces the physical properties and mechanical strength of composites (Seelenbinder 2011). Excessive heating can result from engine overheating or engine fire. Most damage is internal, usually consisting of matrix cracking, which is not easy to detect on the outside of the specimens. This kind of impact damage was observed in composite airframes during flight operations and through the dropping of hand tools during maintenance (Anefaie et al. 2009). Impacts on laminate composites most commonly result in damage such as matrix cracking, fibre fracture, fibre pull-out and delamination (Zelepugin et al. 2018). Stiffness, structural integrity and toughness are greatly reduced by impact loads, reducing the overall performance of the composite structure. In order to understand the above problem, study is focused on material thickness, shape and composition. Shapes of interest are flat, curved, serrated, irregular and circular, and boundary conditions are free, simply supported and clamped (Backman and Goldsmith 1978).

4.3 CLASSIFICATION OF IMPACTS

Impacts can be classified as high or low velocity.

4.3.1 High-Velocity Impacts

Because of the high strain rate, high-velocity impacts are subject to reinforcement, propagation of waves, material stiffness, strength and fracture energy (Hoo Fatt et al. 2010). The effects of impacts on polymer composite materials are influenced by many parameters, namely type, architecture, interaction of reinforcement, thickness, matrix system, projectile geometry and mass (Alireza Sabet et al. 2011). The impact test result is defined as the energy absorbed by the specimen without fracture and is used to find the toughness of a material (Jacob et al. 2002). A highly stressed and greatly overbalanced material can resist a greater impact and is more durable. For example, under high-velocity impact, interlinked fabric composites have higher fracture toughness than simplex laminates (Kim and Sham 2000). Two types of single-stage gun, powder and gas, are used in high-velocity impact testing. In the powder gun test, normal gun powder is the propellant, giving a maximum projectile velocity of 2.0–2.2 km/s. Compressed helium or hydrogen is used as the propellant in a single-stage gas gun (Sultan et al. 2012). Gas gun technologies were not available in the 1950s, but were subsequently developed by US and European government laboratories to provide higher projectile velocities for defence work. High-velocity impact tests have employed flat-faced, complete hemispherical, conical-tip and truncated-cone-nose projectiles 2 to 3mm thick. (Aslan et al. 2003; Raju et al. 1998). Figure 4.1 shows the high-velocity single-stage gas gun type of impact testing device.

Sultan et al. (2012) conducted high-velocity impact tests using a single-stage gas gun. The test rig consisted of a pressure supply unit, a fire machinery unit, a discharge unit, a gas chamber and velocity measurement units. The gun gave a projectile

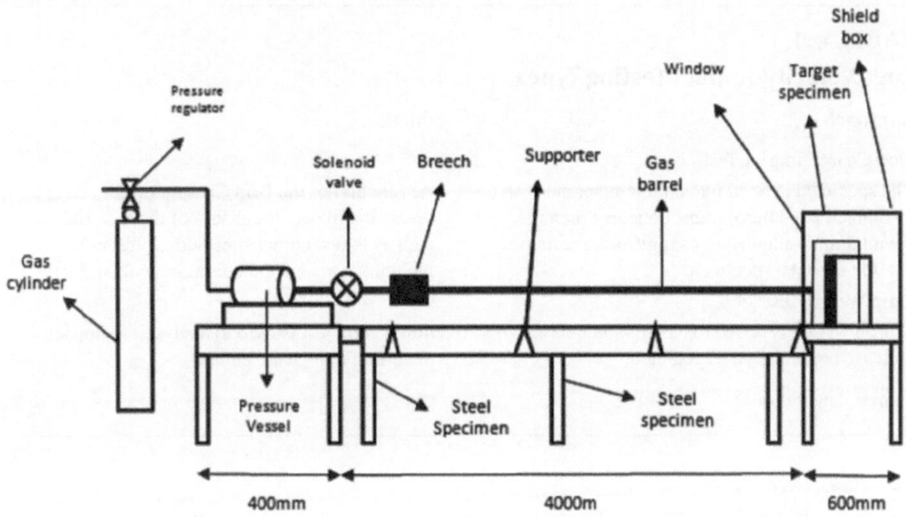

FIGURE 4.1 High-velocity gas gun-based impact testing device (Sultan et al. 2012).

velocity less than 700 m/s and pressure of 150 bars to project the discharge at a reservoir tank. It was concluded that the projectile head did not penetrate any of the specimens, regardless of thickness or velocity. This happened to all fibre-glass specimens of two different types. Microscopically, two types of surface damage areas were investigated: front and back faces.

Mines et al. (1999) measured the high-velocity impact perforation of woven and laminated composites at 570 m/s velocity. Cone, flat and hemispherical projectile nose geometries were used and it was reported that all three configurations behaved similarly.

Ganesh Babu et al. (2006) carried out high-velocity impact tests on unidirectional glass-fibre/epoxy composite plates with a high mass projectile weighing around 550g and using three nose geometries: conical with sharp, round and flat tips. They reported that nose geometry had less influence on energy absorption. The impact velocity exerts a greater effect on thin sections than on thicker sections. When polymer composite materials which are made from various reinforcements are impacted by sharp-tip projectiles at relatively high velocity, one area receives less absorption response (Naik et al. 2006).

4.3.2 Low-Velocity Impacts

A low-velocity impact does not always break the composite. They can also take place during configuration and manufacturing (Shivakumar et al. 1985). Low-velocity impact takes place around 1–10 m/s. The consequences depend on the impact velocity, material characteristics, projectile mass and stiffness. Low-velocity impact is found quite frequently on composite applications such as aircraft components. Aircraft components are exposed to unexpected impact loads due to ground-level procedures and turbulence in flight. A dropped tool impact is of high mass and

TABLE 4.1
Low-Velocity Impact Testing Types

Approach	Function
Izod Charpy Impact Test	
The specimen is fixed into a large machine with a weighted pendulum. Impact occurs when the pendulum is allowed to swing from a certain height onto the specimen.	The function of the Izod Charpy impact test is to assess the impact toughness of the material as well as to test composites with different layers, including twisted and unidirectional laminates.
Drop Weight Test	
A drop weight is raised to a certain height and released to impact the specimen.	Drop weight test is used to analyse the impact pattern of composite laminas.

Source: Duell (2004)

low velocity. The resulting damage takes the form of barely visible delamination (Lopes et al. 2009). In low-velocity impact, delamination and the matrix damage area increase with the impact energy. Less delamination and fewer matrix cracks were observed in tougher materials which have a higher impact resistance than brittle materials. Generally, delamination occurs between the different orientations and mismatched angles of joints (typically 0°/90° or 45°/–45° interface) (Rajbhandari et al. 2002). Rajesh Mathivanan and Jerald (2010) studied the effects of lamination type and the extent of damage with different impact velocities. Marimuthu et al. (2016) analyzed the response of fibre metal laminates to the residual velocity of the impactor. The effects of different failure criteria on the dynamic progressive failure properties of carbon-fibre composite laminates were investigated by Liu et al. (2016). Several different low-velocity impact testing techniques are used for composites. Table 4.1 describes two low-velocity impact tests used in research studies.

4.4 PREDICTION OF IMPACT EFFECTS ON COMPOSITE MATERIALS

Impact force is the shock or high impact exerted on samples in a short period of time. The effect of impact force on composite materials is found to be dependent on the relative impact velocity, being higher for higher velocities. In collisions at normal speeds, one of the impacting materials will absorb most of the impact force. Part of the energy will create deformation or more extreme damage and the remainder will be dissipated in the form of heat or sound energy. The SI unit of energy is the joule (Frederick and Hagy 1986). The measured strength of a structural component made from composite material is affected by several factors, such as the properties of composite material, test condition, geometry and dimensions of the component (Lu and Tongxi 2003). Before considering the impact behaviour of composite materials, it is useful to consider the quality of composite materials.

Modern polymer composites are based on carbon, glass, ceramic, or polymer fibres and have outstanding properties such as high strength, stiffness and low density. The properties of the embedding matrix have an important role in composite behaviour such as transferring stress, protecting the fibres and crack growth (Gholizadeh 2019).

4.4.1 Prediction of Impact Effects

High-velocity impact usually causes more damage and larger bending than low-velocity impact. Low-velocity impact is localized around the immediate area of contact, whereas high-velocity impact results in deformation or damage over a larger area. Two criteria are used to differentiate the velocity regimes. One is based on structural deformation, while the other is based on structural response (Chai and Zhu 2011).

Finite element analysis and experimental measurement are the common methods to predict impact response in all categories of impact. Model superposition and wave controlled are predictive methods for small mass-impact response, and medium mass impact may be predicted using model superposition. A spring mass model, energy balance model and boundary-controlled impact response are all predictive methods for large mass impact.

4.4.2 Impact Responses

Recently there has been considerable research into the inherent properties of composite materials under impact response. Elastic waves propagate from the point of impact into the materials. Energy dissipation and material damping both weaken the elastic wave propagation. Impact duration plays a key role in determining the types of impact response. Responses are found to be reduced for expansion waves when the impact duration is of the same order as the transmission time, as shown in Figure 4.2(a). Flexure and shear waves are found to govern the response for longer impact durations. In Figure 4.2(b) the waves take longer to reach the structural boundary, which results in a quasi-steady-state response. This response happens when deflection and impact load have similar relations.

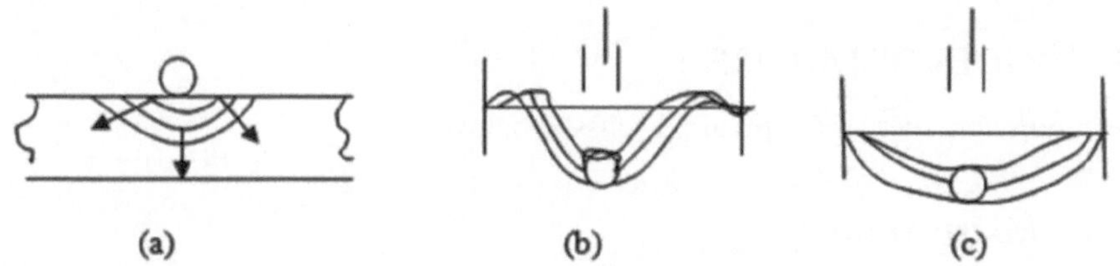

FIGURE 4.2 Different impact regimes and predictive approaches (Chai and Zhu 2011).

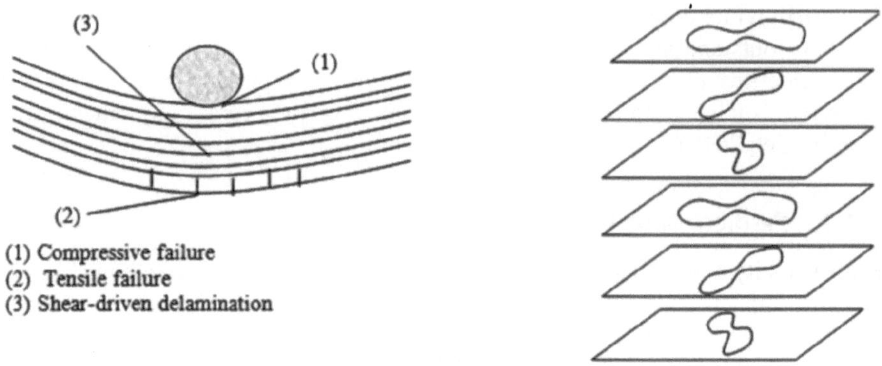

FIGURE 4.3 Different types of impact responses (Zheng 2007).

Figures 4.2(b) and 4.2(c) are associated with the impact responses from runway debris on aircraft structures and dropped weights (such as tools), respectively. In most cases, the impact response is easily detectable (see Figure 4.2(a)). The responses depicted in Figures 4.2(b) and 4.2(c) from wave-controlled and boundary-controlled impacts can cause barely visible impact damage (Zheng 2007; Cantwell and Morton 1990).

The impact responses shown in Figures 4.2(b) and 4.2(c) are related to the shape, dimensions and geometric properties of the specimen. The boundary conditions, impact energy and peak force are the most important parameters to understand the damage mechanisms (Siow and Shim 1998; Rilo and Ferreira 2007). When subjected to impact loads composite materials can absorb and dissipate a large amount of energy through elastic deformation and fracture processes shown in Figure 4.3 (Rilo and Ferreira 2007; Cantwell and Morton 1989). Most of the energy is absorbed through elastic deformation, depending on parameters including interfacial strength, impact velocity and the size of both the component and matrix. In low-velocity impact the incident energy is absorbed by the whole structure, whilst in high-velocity impact most of the energy is dissipated over a small zone near the point of contact (Cantwell and Morton 1989). Because the target response in high-velocity impact is localized, the geometrical effects are very small. In low-velocity impact geometrical effects are greater because the contact time is longer than in high-velocity impacts, and the layer response is also found to be more significant.

4.5 MODES OF FAILURE

The common modes of failure are discussed below.

4.5.1 Matrix Damage

The initial mode of failure is matrix damage, caused by deformation in low-velocity impact. It takes the form of matrix cracking and bonding, due to property dissimilarities between the filament and matrix. The impact response in composite structures is not affected by matrix cracking (Prichard and Hogg 1990). Interlaminar shear and compressive strength properties are decreased in composites which leads to matrix cracks. In unidirectional layers cracks are mostly oriented in planes parallel to the fibre direction.

4.5.2 Delamination

Delamination is particularly important in composite laminates, because this can occur at relatively low loads. Delaminations are highly influenced by flexible stiffness degradation and by buckling failure. Delamination, which is a fracture of weaker interface areas between the composite layers, is found to be the main energy absorption mechanism in polymer-based composite materials. Generally delamination in impacted laminates is driven by the interlaminar shear stresses. The distribution of delaminations in laminated composites is shown in Figure 4.4. It is observed that the major axes of delamination are generally aligned with the fibre direction at the interface (Davies and Olsson 2004).

4.5.3 Fibre Failure

In the fracture process, fibre failure follows matrix cracking and delamination. Fibre failure occurs under the penetrators, due to the local high stress and shear force effects of indentation. Fibre failure is the root cause of complete penetration. The most common fibre failure modes in composite materials are shown in Figure 4.5.

4.5.4 Penetration

Penetration is a macroscopic failure mode, which occurs when fibre failure exceeds a critical extent, allowing complete penetration of the material.

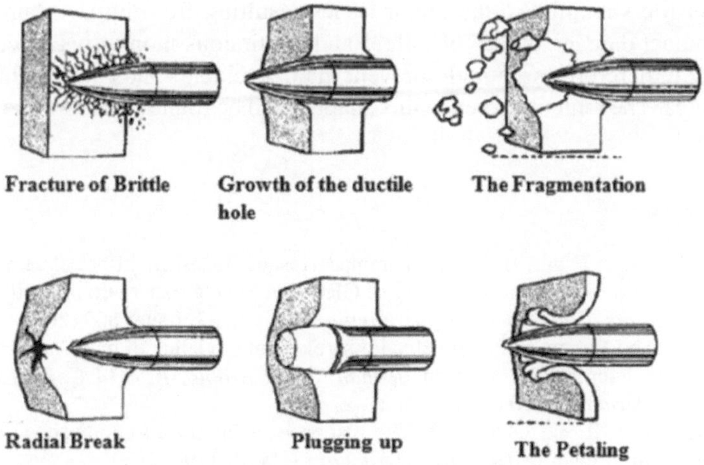

FIGURE 4.4 Distribution of delamination by impact (Zheng 2007).

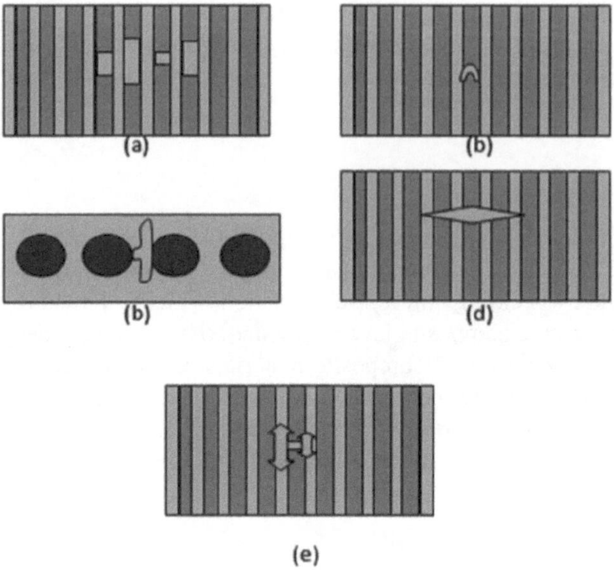

FIGURE 4.5 The common fibre failure modes (Zukas et al. 1990).

4.6 CONCLUSION

Failure of composite materials may be caused by unexpected impacts. Most composites absorb energy through elastic deformation rather than plastic deformation of the composite structure. The structural integrity of composite component structures which are brittle in nature is significantly reduced in a variety of failure modes. Catastrophic failure (or) separation of major structures is not found over prolonged times after impact damage to composites. In design, drastic reduction of compressive and tensile strength are the major issues resulting from impact damage. The study of impact damage modes of failure and continuous monitoring of composite structures are therefore essential to prevent major incidents and catastrophic failure of composites. The topics covered in this chapter will be found useful for the research community.

REFERENCES

Alireza Sabet, Narges Fagih, B. and Mohammad Hosain Beheshty. Effect of reinforcement type on high velocity impact response of GRP plates using a sharp tip projectile. *Tehran, Iran International Journal of Impact Engineering*, Vol. 38, Issues 8–9 (2011): 715–722.

Anefaie, K., M. Abd-Rabou, and N. Bajaba. Finite element modeling of multi-layer composite plates with internal delamination. *Jeddah, Saudi Arabia, ICSV14 Cairns, Australia, Composite Structures*, 90 (1) (2009): 21–27.

Aslan, Z, Karakuzu, R and Okutan, B. The response of laminated composites under low-velocity impact loading. *Composite Structures* (2003): 119–127.

Backman, M.E. and W. Goldsmith. The mechanics of penetration of projectiles into targets. *Int. J. Eng. Sci*, Vol. 16, Issue 1 (1978): 1–99.

Breen, C, Guild, F and Pavier, M. Impact of thick CFRP laminates: the effect of impact velocity. *Composites Part A* (2005): 205–211.

Cantwell, W.J. and J. Morton. Geometrical effects in the low velocity impact response of CFRP. *Composite Structures*, Vol. 12, Issue 1 (1989): 39–59.

Cantwell, W.J. and J. Morton. Impact Perforation of Carbon Fibre Reinforced Plastic. *Composites Science and Technology*, Vol. 38, Issue 2 (1990): 119–141.

Chai, G.B. and S. Zhu. A review of low-velocity impact on sandwich structures. *Proceedings of the Institution of Mechanical Engineers, Journal of Materials: Design and Applications* (2011): 207–230.

Davies, G.A.O. and R. Olsson. Impact on Composite Structures. *The Aeronautical Journal*, Vol. 108, No. 1089 (2004): 541–563.

Duell, Joshua M. Impact testing of advanced composites. *Advanced Topics in Characterization of Composites*, Trafford Publishing (2004): 97–112.

Frederick, E.C. and J.L. Hagy. Factors affecting peak vertical ground reaction forces in running. *International Journal of Sport Biomechanics*, Vol. 2, Issue 1 (1986): 41–49.

Ganesh Babu, M., R. Velmurugan, and N.K. Gupta. Energy absorption and ballistic limit of targets struck by heavy projectile. *Latin American Journal of Solids Structures*, Vol. 3 (2006): 21–39.

Gholizadeh, S. The influence of varying projectile mass on the impact response of CFRP. *International Journal of Mechanical and Production Engineering*, Vol. 7, Issue 3 (2019).

Hawyes, V.J., P.T. Curtis, and C. Soutis. Effect of impact on the compressive response of composite laminates. *Composites Part A—Applied Sciences*, 32 (2001): 1263–1270.

Hoo Fatt, Michelle S. and Dushyanth Sirivolu. A wave propagation model for the high velocity impact response of a composite sandwich panel. *USA International Journal of Impact Engineering*, Vol. 37, Issue 2 (2010): 117–130.

Ibekwe, S.I. et al. Impact and post impact response of laminated beams at low temperatures. *Composite Structure*, 79 (2007): 12–17.

Kim, J.K. and M.L. Sham. Impact and delamination failure of woven-fabric composites. *Composite Science and Technology*, Vol. 60, Issue 5 (2000): 745–761.

Liu, P.F., et al. Finite element analysis of dynamic progressive failure of carbon fiber composite laminates under low velocity impact. *Composite Structures*, Vol 149 (2016): 408–422.

Lopes, C.S., et al. *Simulation of low-velocity impact damage on composite laminates. 50th AIAA/ASME/ASCE/AHS/ASC Structures, Structural Dynamics and Materials Conference* (2009).

Lu, G. and Y.Y. Tongxi. *Energy Absorption of Structures and Materials. Woodhead Publishing* (2003).

Marimuthu, P. et al. Numerical simulation of low velocity impact analysis of fiber metal laminates. *Mechanics and Mechanical Engineering*Vol. 20, No. 4 (2016): 515–530.

Mines, R.A.W., A.M. Roach, and N. Jones. High velocity perforation behaviour of polymer composite laminates. *International Journal of Impact Engineering*, Vol. 22, Issue 6 (1999): 561–588.

Naik, N.K., P. Shrirao, and B.C.K. Reddy. Ballistic impact behavior of woven fabric composites: formulation. *International Journal of Impact Engineering*, Vol. 32, Issue 9 (2006): 1521–1552.

Prichard, J.C. and Hogg, P.J. The role of impact damage in post-impact compression testing. *Composites* (1990): 503–511.

Rajbhandari, S.P., et al. *An approach to modelling and predicting impact damage in composite structures. ICAS CONGRESS* (2002).

Rajesh Mathivanan, N. and J. Jerald. Experimental investigation of low-velocity impact characteristics of woven glass fiber epoxy matrix composite laminates of EP3 grade. *Materials & Design*, Vol. 31, Issue 9 (2010): 4553–4560.

Raju, B.B., Liu, D. and Dang, X. *Thickness effects on impact response of composite laminates. American Society for Composites, 13th Annual Technical Conference on Composite Materials*, Baltimore, MD (1998).

Razali, N. et al. Impact damage on composite structures – A review. *The International Journal of Engineering And Science (IJES)*, Volume 3, Issue 7 (2014): 8–20.

Rilo, N.F. and L.M.S. Ferreira. Experimental study of low-velocity impacts on glass-epoxy laminated composite plates. *Portugal*, Vol. 4, Issue 3 (2007): 291–300.

Safri, S.N.A., M.T.H. Sultan, N. Yidris, and F. Mustapha. Low velocity and high velocity impact test on composite materials – A review. *The International Journal of Engineering and Science (IJES)* Vol 3, Issue 9 (2014): 50–60.

Seelenbinder, J. Composite heat damage measurement using the handheld Agilent 4100 ExoScan FTIR, Easy, non-destructive analysis of large parts. Agilent Technologies, Connecticut, USA (2011): 5990–7792.

Shivakumar, K. N., Elber, W. and Illg, W. Prediction of low velocity impact damage in thin circular laminates. *American Institute of Aeronautics and Astronautics Journal* (1985): 442–449.

Siow, Y.P. and V.P.W. Shim. An experimental study of low velocity impact damage in woven fibre composites. *Journal of Composite Materials*, Vol. 32, No. 12 (1998): 1178–1202.

Sultan, M.T.H. et al. High velocity impact damage analysis for glass epoxy – Laminated plates. *Advanced Materials Research*, Vols. 399–401 (2012): 2318–2328.

Zelepugin, S.A., A. A. Popov and D. V. Yanov. Failure of the laminate composites under impact loading. *6th International Congress "Energy Fluxes and Radiation Effects*, IOP Conf. Series: *Journal of Physics: Conf. Series*, Vol. 1742–6596 (2018).

Zhang, X.X., G. Ruiz and R. C. Yu. A new drop weight impact machine for studying the fracture behaviour of structural concrete. *Transactions on the Built Environment*, Vol 98 (2008): 1743–3509.

Zheng, D.H. Low Velocity impact analysis of composite laminated plates. phd thesis, The Graduate Faculty of the University of Akron (2007).

Zhou, G. Damage mechanisms in composite laminates impacted by a flat-ended impactor. *Composites Science and Technology*, 54 (1995): 267–273.

Zukas, J.A. et al. *Impact Dynamics*. John Wiley & Sons, New York (1990).

5 Challenges of Adhesively Bonded Joints and Their Advantages over Mechanical Fastening

B. N. V. S. Ganesh Gupta K
National Institute of Technology, Rourkela, Odisha, India

Kishore Kumar Mahato
Vellore Institute of Technology, Vellore, Tamilnadu, India

Rajesh Kumar Prusty and Bankim Chandra Ray
National Institute of Technology, Rourkela, Odisha, India

CONTENTS

DOI: 10.1201/9781003128861-5

5.1 INTRODUCTION

Fibre-reinforced polymer (FRP) composite materials are recognized as valuable materials in structural applications due to their superior performance over traditional materials such as aluminium and steel. Composite materials are generally produced by combining two or more different materials (as shown in Figure 5.1) having different physical or chemical properties, which results in better performance than the

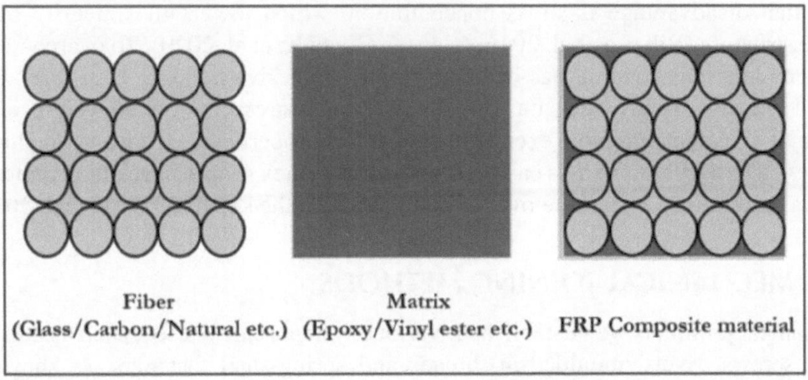

FIGURE 5.1 Schematic representation of a composite material.

individual materials used singly. Moreover, multilayer fibre-reinforced composite materials offer a high strength-to-weight ratio, low density, excellent corrosion resistance and improved fatigue and wear properties. They can also absorb higher energy than traditional materials in an impact (Prusty et al. 2015; Ganesh Gupta K et al. 2020). Design engineers also select composite materials over traditional materials in application areas such as aerospace, railways, civil engineering, automotives and energy due to their flexible production process, long-term stability, lower maintenance costs and high chemical resistance.

5.2 SIGNIFICANCE OF MULTI MATERIAL JOINTS

FRP composite materials exhibit limited performance when exposed to harsh environments like seawater, strong chemicals, moisture and humidity cycles and service temperature variations. The durability of FRP composite materials is questionable for long-term applications when exposed to such environmental challenges (Prusty et al. 2017). High-performance materials are attractive, meeting the global requirements for strength, economy, safety and environmental impact, especially in automotive and structural applications. By combining different materials these new requirements may be achieved which results in lower carbon emissions and structural weight, and improvements in durability even under harsh environmental conditions (Ray and Rathore 2014). Most assemblies combine two or more components. Combining different materials (e.g., composite and steel/aluminium) can compensate for the inherent weaknesses of the individual materials. The combined effect of properties obtained by adhesively bonding FRP composite material and conventional material (metal) has gained high popularity worldwide because it can give improved performance without compromising safety. These materials can potentially replace conventional materials in structural applications. They can yield efficient, safe and economic structures of reduced weight and high durability. Numerous techniques are available to join different materials (Ganesh Gupta K et al. 2021). One of the most utilized is mechanical fasteners, which come with disadvantages such as increased structure weight and the possibility of galvanic corrosion. However, the most important disadvantage is stress concentration, which lowers the integrity of the whole structure (Ribeiro et al. 2016; (Ganesh Gupta K, et al. 2020). To overcome this problem adhesive bonding has been developed. Adhesive bonding is now in widespread use, as it can provide a uniform distribution of stress, low added weight, elimination of galvanic corrosion, excellent mechanical properties, low density and higher stiffness. Dissimilar materials can be joined and complex shapes have much improved flexural and fatigue resistance over bolted joints (Budhea et al. 2017; Wahab 2016).

5.3 MECHANICAL JOINING METHODS

Mechanical joining is well established. Various systems can be used, such as nuts and bolts, screws, rivets, moulded-in threads and spring-steel fasteners, as shown in Figure 5.2 (Troughton 2008; Lathabai 2011). Mechanical methods are suitable for joining metal parts, composite parts and multi-material combinations, i.e., metal to composite.

5.3.1 Nuts and Bolts

Assembly using nuts and bolts involves drilling holes in the mating parts at the required position using an appropriate drill and clamp (Küçükoğlu and Karpat 2017). Bolts may join the parts in combination with a nut or by entering a threaded hole made in one part, and one or more washers are generally used. Three parts of the bolt play essential roles: the head, the shank and the thread. The bolt passes through the drilled holes in the parts with the bolt head and nut on opposite faces of the mating parts (Messler, 2004). In this joining method, the nut applies an axial holding force, and the shank (the unthreaded part of the bolt) works as a dowel.

The bolt shank length must be equal to the total thickness of the joint. Too long a bolt thread will extend beyond the nut. It may interfere with the movement of nearby parts and adds unnecessary weight. If the shank is too short, the thread may not extend out of the bolt hole far enough for the nut to be securely fastened. The shank (un-threaded) length is also called grip length. The internal diameter of washers is similar to the shank diameter of the bolt. During assembly washers are generally placed under the head of the bolt and under the nut to hold and support the system.

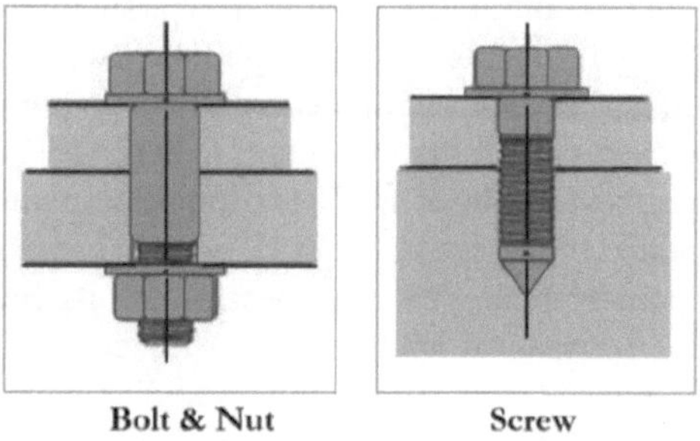

FIGURE 5.2 Schematic representation of mechanical fastening.

Numerous types of bolts are available for different applications, including hex bolts, anchor bolts, structural bolts, carriage bolts, etc.

Some examples of bolt use are:

- Hex bolts, entirely or partially threaded, with hexagonal heads are extremely common in all forms of engineering.
- Machine bolts, having square heads and a fully threaded shank, are suitable for joining wood and metalwork.
- U-bolts' primary application is to hold or support pipes. U-bolts have a threaded portion at both ends with the central U unthreaded.
- Structural bolts have a standard hex head of reduced height, and are much used in structural applications.

5.3.2 Screws

Screw fasteners are the most common threaded fasteners. They are widely accepted in the automotive and aircraft industries (Troughton, 2008). Various types are available, including wood, structural, machined, threaded, sheet metal, self-tapping and socket-head. There are many different head styles: flat, oval, pan, truss, hex, round, socket-head, button, hex washer, slotted hex washer. For example, round-head screws are widely used in high-stress aircraft parts, whereas pan and truss screws are used for general applications.

Structural screws: These are simple in design and appropriate for structural applications. They are made of heat-treated alloy steel.

Self-tapping screws: Self-tapping screws produce internal threads as they are turned into the hole while assembling the parts.

Machine screws: Machine screws are most useful to hold or join heavy metal components.

5.3.3 Rivets

Rivets are a fast and flexible assembly process. They are globally accepted and widely used in aircraft and automotive applications for joining similar and dissimilar materials (Lathabai, 2011). Design engineers recommend that the rivet head diameter must be three times that of the shank to minimize or eliminate stress concentration (Wahab 2016). Unlike the various forms of screwed joint, riveting is a permanent fastening method.

Solid and blind rivets are the most favourable. Solid rivets are simple in design, most reliable, and commonly made of aluminium, steel, copper or brass. A solid rivet has a head at one end, and a solid cylindrical shank. Prior drilling is required as for other mechanical fastening methods. The solid rivet is pushed into the mating parts' aligned holes. Then external force is applied, using a rivet gun, to the exposed end of the shank. The shank end deforms to create a second head, thus joining the materials together.

Blind rivets have a hollow shank through which passes a mandrel. The blind rivet is inserted into the aligned holes of the mating parts and the rivet mandrel is withdrawn from the head end using a special tool. This process expands the rivet's shank, thus forming the closing head. On reaching a predetermined breakpoint the exposed end of the mandrel breaks off and is discarded (Messler 2004). Various types of blind rivet are available, namely open end, closed end, multi-grip, peel type, avinox and two-piece mate.

5.3.4 Moulded-In Threads

Moulded-in threads are one of the most versatile joining methods. A thread is moulded into both mating parts. Moulded-in thread fasteners are quite common in applications where assembly and disassembly process is infrequent (Wahab, 2016). This process is most suitable for joining thermoplastic to thermoplastic components (e.g., a water bottle and its cap) but exhibits poor performance when joining dissimilar materials like plastic to metal. The difference in thermal expansion leads to higher stress concentration in the plastic component.

5.4 ADHESIVE BONDED JOINTS

5.4.1 Adhesive

Adhesives can provide permanent and strong bonding between substrates made of similar or dissimilar materials (Budhea et al., 2017; Wahab, 2016). An adhesive or sealant must meet three fundamental requirements to adhere to and/or seal the components. First, the material has, at some stage, to be in a fluid state such that it can promptly spread over and wet the substrate. Second, the material must solidify. Finally, the material must resist challenges that it will see during manufacturing (Petrie, 2007).

Adhesives are classified into two types: structural adhesives, which possess high shear and tensile strength; and non-structural adhesives which have lower strength and durability. Structural adhesives are mostly used in critical assemblies (aerospace, civil construction, automotive, health and electronic sectors), where the material must withstand different environments and provide high load-bearing capacity. Non-structural adhesives are generally used in commercial applications like joining sensitive films to substrates.

5.4.2 Adhesion

In adhesively bonded joints, both adhesion and cohesion forces play an important role in joint behaviour. Adhesion is the tendency of two or more dissimilar molecules to bond with each other, whereas the force of attraction between similar molecules is known as cohesion, also called the internal strength of the material (Budhea et al., 2017; Ebnesajjad, 2009). The interfacial forces holding the two adherends together may arise from van der Waals forces, adsorption, interdiffusion, chemical composition, mechanical interlocking and electrostatic attraction.

FIGURE 5.3 Schematic representation of adhesively bonded single lap joint.

5.4.3 Adherend

An adherend is also called a substrate, where the adherend surface adheres/joins to another adherend by adhesive action. Figure 5.3 shows schematically an adhesively bonded single lap joint (SLJ) where a shim has been used to prevent misalignment in testing conditions.

5.4.4 Adhesive Types

Polymers are classified into thermoplastics and thermosets (Petrie, 2007). Thermoplastic materials are reversible and physically solid at room temperature. They become flowable at a specific elevated temperature, which is above the melting point for semicrystalline thermoplastics and above the glass transition temperature for amorphous thermoplastics material. They resolidify upon cooling. Thermoplastics include polyethylene, polypropylene, polystyrene and polyvinylchloride.

Thermosetting polymers are liquid at room temperature and become solid when heat and pressure are applied. These changes are irreversible. When thermosetting materials are heated beyond their glass transition temperature (where the material changes from a solid glassy state to a soft rubbery state) they begin to degrade and will be rapidly decomposed beyond their decomposition temperature (Petrie, 2007; Ebnesajjad, 2009). The performance of thermosetting materials is strongly dependent on the cross-linking of the polymer chains. Thermosets such as epoxy, vinyl ester and phenolic have found widespread application across the globe, especially in structural industries, due to their flexibility, higher chemical resistance, ease of fabrication and superior performance.

Epoxy polymers are extensively used in structural components due to their physical, mechanical, thermal, physicochemical properties and high dimensional stability (Banea and Da Silva 2009). Epoxy polymers are used as engineering adhesives and matrices for fibre-reinforced composite materials. Epoxy polymers usually require a high level of cross-linking, which is influenced by the curing process and the stoichiometric ratio of resin to curing agent. When cured, epoxy polymers have high cross-link density, exhibiting excellent adhesion, good performance at elevated temperature, good thermal stability, creep resistance and a relatively high modulus. Vinyl ester is manufactured by reacting epoxy resin with methyl acrylic acid. A reactive diluent such as styrene is used in proportions between 30% and 50%. This acts as a co-monomer and the curing of vinyl ester proceeds through free radical copolymerization with styrene (Ganesh Gupta K et al. 2020) Vinyl ester provides similar mechanical properties to epoxy and polyester since it is an unsaturated ester of epoxy resin. Vinyl ester uses the same processing and curing techniques as the polyesters.

5.4.5 Advantages of Adhesive Joining

- Ease of fabrication
- Freedom of design and material selection
- Uniform stress distribution
- Negligible weight addition
- Suitable for complex shapes
- Suitable for thin material
- Suitable for adhering to large surface areas
- Act as a tolerance compensation
- High corrosion and chemical resistance
- No galvanic corrosion
- High flexibility
- Act as noise insulation
- Better damping properties
- Improved product durability and reliability
- Enhanced product aesthetics
- Thermal and electrical insulation
- Lower production cost (Petrie 2007; Wu et al. 2020).

5.4.6 Adhesive Disadvantages

- Surface treatment prior to bonding is essential
- Limited thermal loads
- Additional heat energy is required for curing
- Specific production process
- Difficult in quality inspection (Petrie, 2007; Wu et al.,2020).

5.5 TYPES OF ADHESIVE BONDED JOINTS

An adhesive bond between two adherends is formed when an adhesive medium, applied between the adherends, changes from the liquid state into solid form by applying heat and pressure. Molecular-scale interactions (Van der Waals, hydrogen bonds and covalent bonds) between adhesive and adherend surfaces create a permanent bond between two adherends (Wahab 2016). Strong chemical bonding at the interface provides effective stress transfer between the adherends. Adhesively bonded joints are classified into three types: co-curing, co-bonding and secondary bonding (Budhea et al., 2017; Fuertes et al., 2015).

5.5.1 Co-Curing

Co-curing is a combined manufacturing process for developing structural materials. Uncured composite material adherends and adhesives are cured simultaneously. The most common bonding mechanism in the co-curing process is chemical

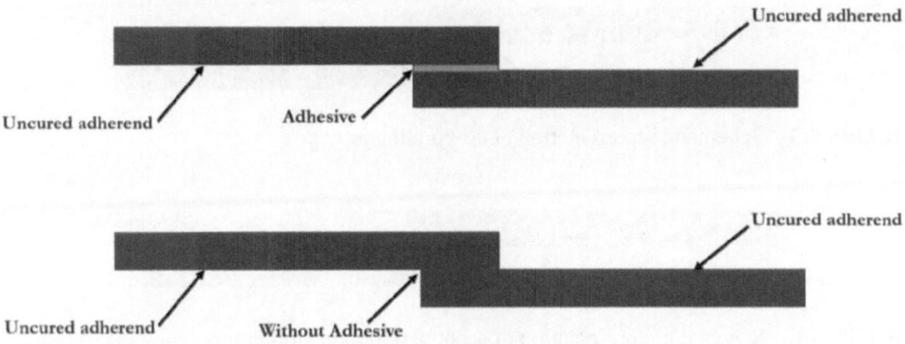

FIGURE 5.4 Schematic representation of co-curing joints.

cross-linking (Fuertes et al., 2015). The co-curing process can in some cases occur without adhesive for joining two or more uncured adherends (as shown in Figure 5.4).

Liquid composite moulding processes (LCM) techniques are also called co-curing processes. Variants include vacuum-assisted resin infusion moulding (VARIM), vacuum-assisted resin transfer moulding (VARTM), Seemann composites resin infusion moulding process (SCRIMP) and vacuum-assisted resin infusion process (VARI). All these approaches involve the impregnation of a dry reinforcement by liquid thermoset resin driven under vacuum. The co-curing technique is mostly used in the automotive and aircraft industries to assemble a large number of minor parts into a single structural unit, and is widely accepted in the repair of composite structures (Fuertes et al., 2015; Streitferdt et al., 2017). The combined process reduces the processing time and curing time, which may lower the final cost of the structural unit. An example is given below.

VARIM is a composite material fabrication process used to produce high-quality, large-scale components. Dry preform fabrics, release film and peel plies are placed on an open mould, and a plastic vacuum bag closes the top side of the mould. A vacuum pump then evacuates the air. Liquid resin mixed with a hardener is drawn from an external reservoir into the component system by the vacuum. The resin with the hardener infuses into the preform until impregnation is complete. Finally, curing and de-moulding steps finish the product.

5.5.2 Co-Bonding

In the co-bonding fabrication technique, one uncured composite material is bonded with another pre-cured composite material with an adhesive film medium. The process of co-bonding is shown in Figure 5.5. Prior surface treatment is required for the pre-cured adherend to remove unwanted contaminants, glossy nature and provide suitable surface roughness for bonding. This process is also called the intermediate bonding process. The two different joining mechanisms take place between co-bonded structural components. Chemical crosslinking takes place between the uncured adherend and the adhesive film, whilst an adhesion mechanism takes place between the adhesive film and the pre-cured adherend (Fuertes et al., 2015).

FIGURE 5.5 Schematic representation of co-bonding joints.

FIGURE 5.6 Schematic representation of secondary bonded joints.

5.5.3 Secondary Bonding

The secondary bonding process is the most common joining process in various structural applications. Where the adhesive film or adhesive paste is used to join two or more different adherends the secondary process requires two steps. (Fuertes et al., 2015; Streitferdt et al., 2017). The adhesive medium is placed at the interface of the pre-cured composite materials as shown in Figure 5.6. Adhesion occurs between the pre-cured adherends and the adhesive, giving strength to the joint. However, prior to joining, surface treatment of the adherend materials is necessary to enhance the surface morphology of the adherend surfaces at the bonding area. Treatment also improves the wettability between the adherends and adhesive medium. Secondary bonding is mostly used to create high strength and efficient structures, and this method is well suited to multi-material joints and complex structures (Wu et al. 2020).

5.6 BONDED JOINT DESIGN AND IMPORTANCE OF PROCESS PARAMETERS

For developing strong and long durable structures, substrate material selection is a primary factor. Bonding properties also play an important part along with the physical appearance, mechanical strength, cost and production process in the overall performance of the structures (LeBacq et al., 2002). Adhesive joining technology is compatible with metals, glass, ceramics, wood, plastic, paper, and FRP composites.

Adhesively (identical substrates and dissimilar) bonded joints must follow design rules to meet the global requirement and needs of industry (Petrie, 2007; Jensen et al., 2016). These rules concern surface treatment, adherend geometry, adhesive thickness, bonding area, load limit, safety factor, expansion, creep-relaxation, zero-error production and testing/fault analysis.

5.7 SURFACE TREATMENT

Surface treatment decreases water contact angle, increases surface tension, and as a result increases bond strength. The primary aim of surface treatment is to increase the adherend's surface energy as much as possible, thus improving the bond

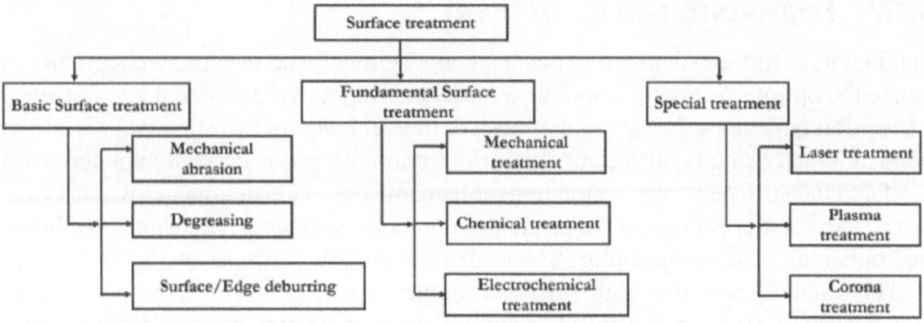

FIGURE 5.7 Schematic representation of surface treatment methods.

efficiency (Streitferdt et al., 2017; Oldewurtel, 2019). Various types of surface treatment are available (Figure 5.7), and the actual method to follow depends on the type of adherend/substrate. The factors involved in selection of surface treatment for adherends are material type, adherend surface condition, adherend geometry, manufacturing method and assembling technique (LeBacq et al. 2002; Oldewurtel 2019).

5.7.1 Basic Surface Treatments

Basic surface treatments such as mechanical abrasion, peel ply, degassing and surface/edge deburring are the most common techniques used to create the necessary surface texture on adherend surfaces for better adhesion (Streitferdt et al., 2017; Ebnesajjad, 2009). These remove contamination like dust, grease, sand particles, debris, any oxide layer or rust (for metal) and glossy nature (for FRP composites). For example, peel ply is widely applied to alter the composite material's surface morphology. The peel ply is placed onto the composite adherend and removed prior to bonding. Peel ply, which is suitable for bonded joints, improves roughness of the adherend surface. The most effective type of peel ply depends on material type, adhesive type and other factors (Kanerva and Saarela 2013). The common steps of basic surface treatment are listed below (Banea et al. 2017; Ebnesajjad 2009):

- Clean the adherend surfaces with acetone or other solvent using a lint-free cloth.
- Ensure that all abrasive materials to be used in basic surface treatment, such as grit, sand, silicon carbide paper or brushes, are free from unwanted contaminants.
- Wire brushing and abrasion with silicon-carbide paper roughen the adherend surface, making it suitable for the joining process.
- Clean the adherend surfaces with dry air from an air pump to remove any tiny particles.
- Finally, prior to bonding of the adherends, clean again with acetone or other solvent. Wiping must be in one direction, not back and forth.

5.7.2 Fundamental Surface Treatment

Fundamental surface treatments (mechanical, chemical, and electrochemical) aim to create the optimum adherend surface texture and improve the physical and chemical adsorption between substrate and adhesive, thus enhancing the effectiveness of the bonded joints. Primary surface treatment is commonly and necessarily applied in all fundamental surface treatment methods. Mechanical surface treatment involves various processes, depending on the production process, such as sandblasting, grit blasting, polishing, and shot peening (Saeedikhani et al., 2013; Arenas et al., 2013).

This method provides high surface roughness and surface energy, resulting in improved bond strength. For metal adherends chemical treatments that etch the adherend surface are considered very important. This method improves the wettability and adhesion strength by enhancing surface roughness. Its effectiveness depends strongly on parameters like chemical composition, bath temperature and etching time (Wahab, 2016; Saleema et al., 2012). A form of electrochemical treatment, known as anodisation, is a special process that significantly alters the substrate morphology (Petrie, 2007; Saeedikhani et al., 2013). This method creates an anodic film of oxide, typically between 0.1 μm and 5.0 μm thick, that appears as an opaque iridescent film on the aluminium surface. Anodisation greatly improves the performance of the bonded structures, but the resulting adhesion performance is strongly influenced by electrolyte concentration, applied voltage, treatment temperature and immersion time.

5.7.3 Special Surface Treatment

Special surface treatments (laser, plasma and corona discharge) create the optimum surface roughness on the composite adherend surface. The physicochemical activity of these treatments also improves the bond strength between the adhesive and adherend (Encinas et al., 2014; Li et al., 2009). Corona surface treatment is an electrical discharge method that applies a high frequency and high voltage between two electrodes in atmospheric air. This method creates oxidized chemical groups such as carbonyls, hydroxyls, ethers and esters, and it increases the surface roughness (Ebnesajjad, 2009). The electrical discharge originates from the adhesive surface, and this method is most effective for plastic films.

5.7.4 Quality Assessment of Surface Preparation using Water Break Test

A water break test is an essential testing method to evaluate the quality of the prepared adherend surface (Ebnesajjad, 2009). It depends on the observation that a clean surface will hold a continuous water film rather than a series of isolated droplets. A break in the water film indicates a soiled or contaminated area, which is not suitable for the bonding process.

5.7.5 Adherend Geometry

To achieve high performance in structural joints, stress concentration sites in bonded joints/structures must be eliminated. This can be achieved by choosing the proper

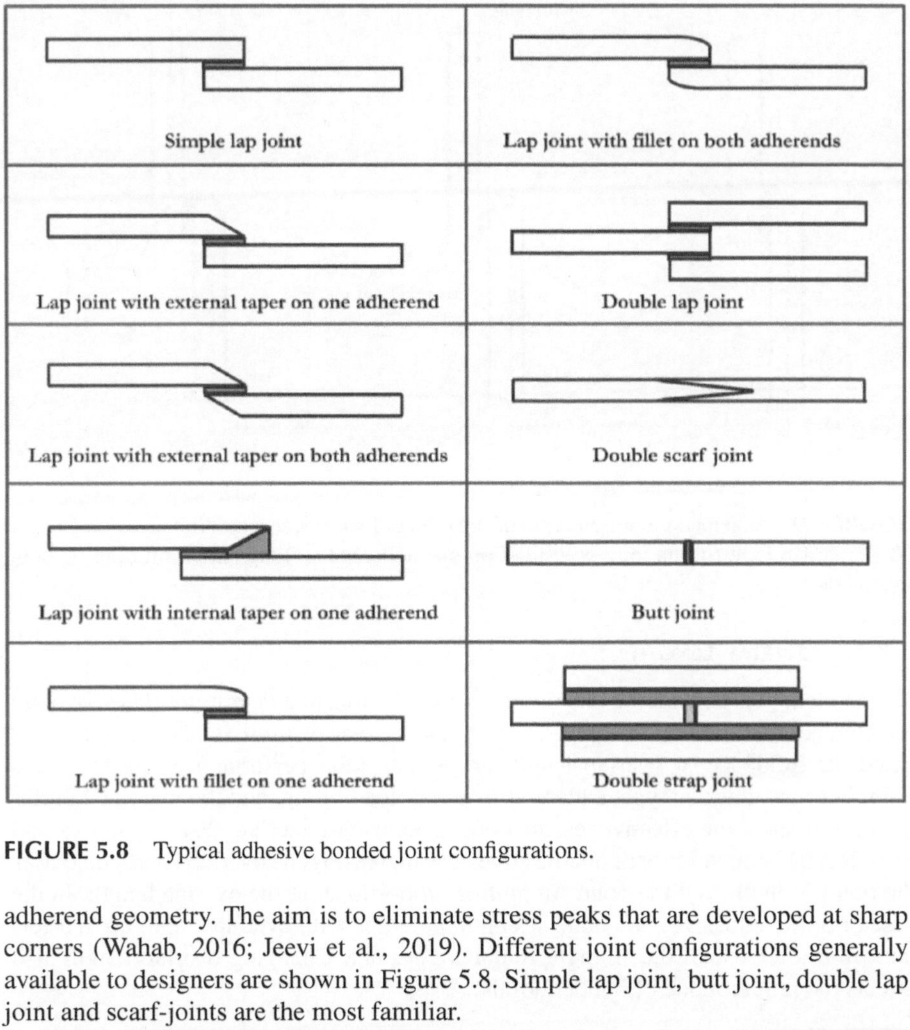

FIGURE 5.8 Typical adhesive bonded joint configurations.

adherend geometry. The aim is to eliminate stress peaks that are developed at sharp corners (Wahab, 2016; Jeevi et al., 2019). Different joint configurations generally available to designers are shown in Figure 5.8. Simple lap joint, butt joint, double lap joint and scarf-joints are the most familiar.

Design engineers focus on improving joint designs to reduce or eliminate stress concentrations. For example, in a simple lap joint, stress concentration is higher at the bonded area due to a sharp joint geometry. A fillet on one of the adherends lowers the stress peaks for the same bonded area. Stress concentration is further decreased if the lap joint has a fillet on both adherends, which is clearly shown in Figure 5.9. The strength of the joint is severely influenced by its design.

5.7.6 Adhesive Thickness

The effectiveness of bonded joints is strongly influenced by adhesive thickness; adhesive thickness is a prime factor for achieving a durable joint (da Silva et al., 2009). Increasing the adhesive thickness increases deformation, resulting in lower strength and higher elongation at fracture. Reducing the thickness has the opposite effect.

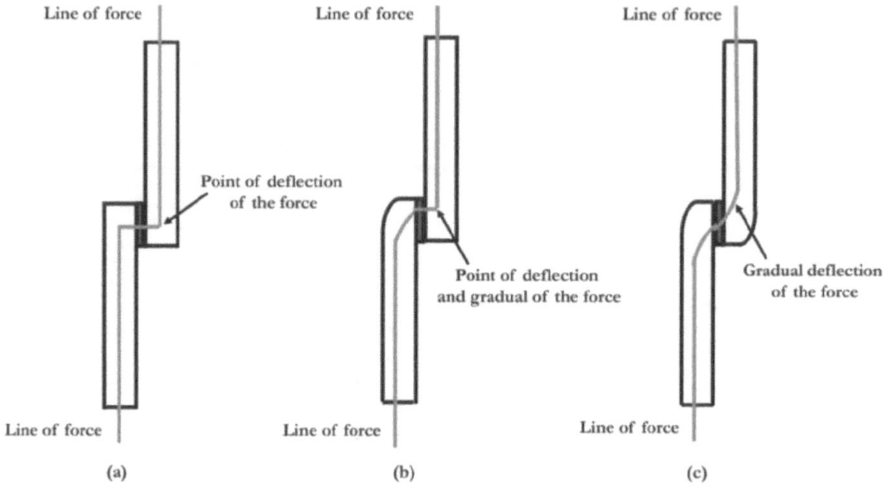

FIGURE 5.9 Schematic representation of the effect of stress concentration in bonded joints (a) simple lap joint (b) lap joint with fillet on one adherends (c) lap joint with fillet on both adherends.

5.7.7 Overlap Length

Overlap length is considered one of the primary factors that determine adhesive bond effectiveness (Wu et al., 2020; Reis et al., 2011). Various researchers have investigated the influence of overlap length on bonded joint performance, but they also point out that adhesive type, adherend material, load rating, and the material combination influence the effectiveness of bonded joints (Reis et al,. 2011; Kanerva and Saarela, 2013). In the case of ductile adhesives increasing the overlap length increases the bond strength, and the bond strength is proportional to the overlap length. In the case of brittle adhesives, the bond strength increases with overlap length up to a certain limit after which the strength remains constant. Changing overlap length also affects the failure mode of bonded joints.

5.7.8 Fabrication of Adhesively Bonded Joints

Before fabrication of bonded joints all the pre-processing steps, i.e., surface treatment, adherend type, adhesive type and adhesive thickness, must be completed (Özes and Nurhan Ne, 2015). When this has been done, an appropriate process for assembling the bonded joints can be selected as shown in Figure 5.10.

5.7.9 The Steps Involved in Bonded Joint Fabrication (Jensen et al. 2016; Ebnesajjad 2009):

- Ensure all safety precautions are in place and the ambient conditions are suitable for the assembly process.
- Confirm that all necessary tools (fixing/positioning aids, shims, spacers) are available and working correctly.

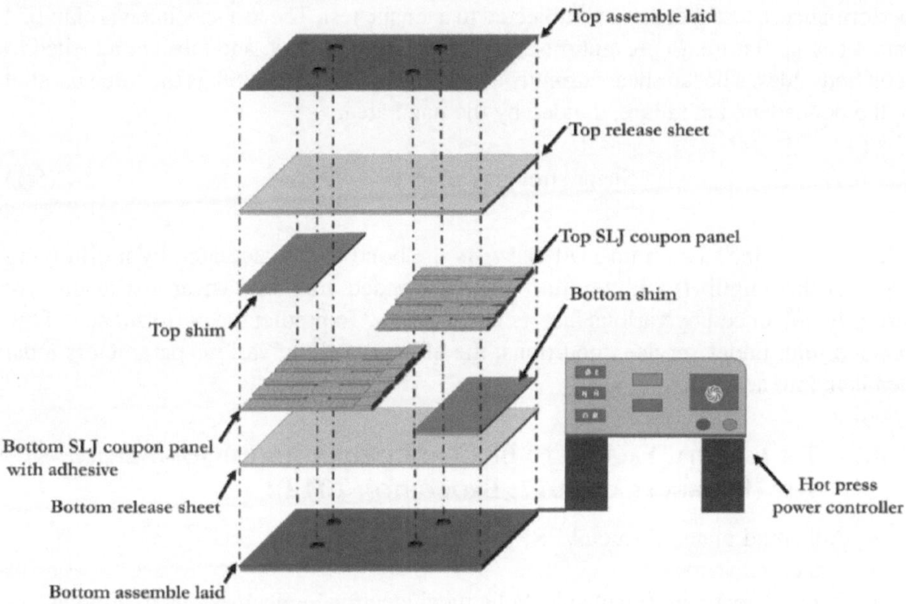

FIGURE 5.10 Schematic representation of fabrication and assembly of single-lap-joints (SLJs).

- Prepare the adherend surfaces, carefully handling the prepared surfaces to avoid contamination.
- Put one adherend onto the assembly bed and use fixing aids to prevent movement.
- Prepare the adhesive according to the instructions then, using clean appropriate tools, apply it to the lower adherend surface over the desired bond area. Then, with the desired overlap, place the upper adherend onto the lower, adhesive-coated, adherend.
- Shims are the most appropriate tool to obtain the required adhesive thickness.
- When the adhesive has achieved sufficient strength for handling, remove the finished joints from the assembly bed.
- Finally cure the joints using the optimum conditions. Curing enhances crosslinking and removes residual stresses.

5.8 MECHANICAL CHARACTERIZATION OF BONDED JOINTS

Mechanical characterization techniques are used to measure the performance of finished adhesively bonded joints.

5.8.1 Lap Shear Tests

The lap shear test is a destructive test method. It is a reliable and popular method for measuring bonded joint performance (Ribeiro et al., 2016; Özes and Nurhan Ne, 2015). In this test, the applied stress acts parallel to the plane of the joint. To assess the joint performance, a test specimen is subjected to a tensile test. The test specimen is clamped into a testing fixture, to prevent misalignment during the test, and tabs are attached to it on both sides. The lap shear strength, according to ASTM D5868, is the force parallel to the bonded area at failure, divided by the bond area.

$$\text{Shear strength}\left(\tau\text{max}\right) = \frac{F_{\text{max}}}{A} \tag{5.1}$$

where F_{max} is the maximum load and A is the bond area, calculated by multiplying together the length (l) with width (b) of the bonded area. Lap shear test results are strongly influenced by various factors listed below. To predict the performance of the bonded joint under service conditions, the effects of these various parameters must be taken into account.

5.8.2 The Essential Factors to Be Considered in the Lap Joint Efficiency Test (Budhea et al. 2017; Broughton 2012):

- Adherend material (metal/FRP)
- Adherend shape
- Surface treatment (mechanical/chemical/electrochemical/special treatments)
- Adhesive type (thermoplastic/thermoset)
- Application of adhesive (mixing paste/film)
- Adhesive thickness
- Variation in the bonded area
- Loading rate
- Curing temperature conditions
- Curing time conditions
- Testing temperature conditions
- Exposure time
- Bonded joints fabrication type
- Environmental aging (humidity, hydrothermal, hygrothermal, UV radiation) conditions

5.8.3 Fatigue Tests

Fatigue tests are essential to select adhesives for aerospace, automotive, civil, etc. structural applications (Broughton, 2012). Test method ASTM D3166 covers the measurement of fatigue strength in shear by tension loading of adhesives on a standard specimen and under specified conditions of preparation, loading and testing. The method may be used for measuring the fatigue properties of adhesives using plastic adherends, provided consideration is given to the thickness of the plastic adherends. Doublers may be required for plastic adherends to prevent bearing failure in the adherends.

5.8.4 Creep Tests

For real-world applications measurement of the creep response of adhesively bonded joints is an essential mechanical test. It is beneficial for understanding the deformation behaviour of bonded joints under different conditions of load and/or temperature (Ebnesajjad, 2009; Broughton, 2012). ASTM D2294 provides two methods of creep measurement: shear by tension or shear by compression.

5.8.5 Post-Failure Analysis

ASTM D5573 lays out a standard practice for classifying failure modes in fibre-reinforced plastic (FRP) joints.

Various failure modes and behaviours of bonded joints at the macroscopic scale are considered, to support failure mode analysis of adhesively bonded joints under various testing conditions (Zhang et al., 2013; Petrie, 2007). Adhesive failure is rupture of the adhesive bond such that separation occurs at the adherend/adhesive interface. Information on interfacial strength is usually obtained from the lap shear test. The term "interface" means the region immediately surrounding the junction between the adherend and adhesive. The material properties in this region tend to differ significantly from the bulk adhesive. Adhesive failure may be caused by poor surface quality and by hydrated adhesive and adhesive chemical bonds. In cohesive failure mode, the adhesive itself fails and is found on both adherends. This type of failure may occur by shear, peel or a combination of shear and peel stresses. Some adhesive joint design errors cause cohesion failures, such as excessive peel stress, inadequate overlap area and voids in adhesives. The typical failure modes of adhesive bonded joints are shown in Figure 5.11.

5.8.6 Differential Scanning Calorimetry (DSC)

DSC is the best tool to determine the glass transition temperature (Tg) of polymeric adhesive materials. Tg is the temperature at which the rigid glassy state of a polymeric material changes to the soft rubber state. It is a critical factor in end-use applications (Ganesh Gupta K et al. 2020). DSC testing parameters include temperature range, heating rate and testing atmosphere.

5.8.7 Dynamic Mechanical Thermal Analysis (DMTA)

DMTA according to ASTM D7028 is the most suitable method for measuring the viscoelastic response of a polymeric adhesive under applied mechanical and thermal loads. Test parameters include temperature range, frequency, atmosphere and heating rate. This test measures the following viscoelastic properties. Storage modulus (E'), loss modulus (E"), and damping tendency (tanδ) of the experimental material. The DMTA test can also provide the glass transition temperature (Tg) using E' and tanδ plots (Prusty et al. 2017). The damping factor tanδ, is a significant parameter in automotive structural applications.

5.8.8 Computational Simulation Studies Related to Adhesives

Theoretical and computational approaches are increasingly important for advanced research in various fields including adhesion science, tribology and rubber

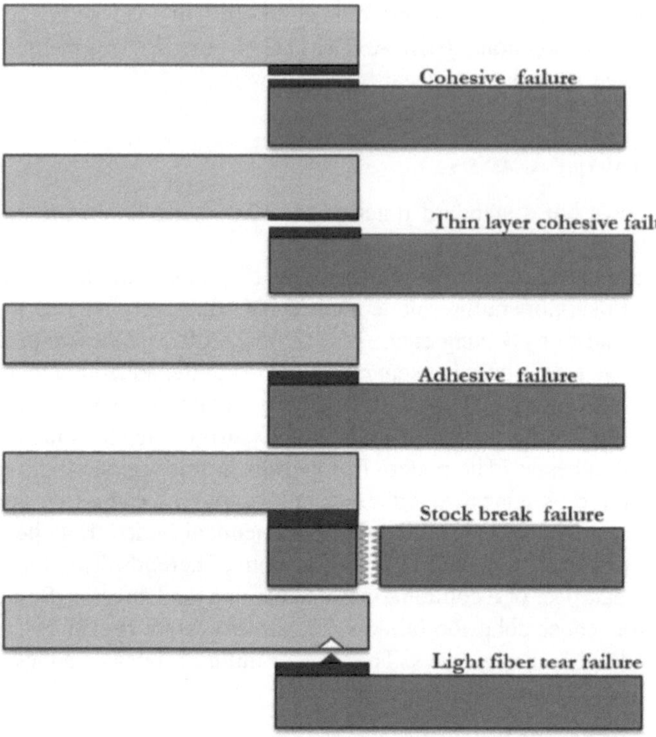

FIGURE 5.11 Schematic representation of failure modes of bonded joints (adapted from Gupta et al., 2020c).

reinforcement by nanoparticles. These computational tools complement experimental investigations. Computational and simulation tools have become essential for designing new materials and predicting their behaviour under various conditions. Theoretical and computational modelling can reduce the cost and duration of experimental measurements. A multiprocessor computer is required, together with codes based on molecular dynamics, quantum mechanics and statistical mechanics. Baggioli et al. (2020) suggested that molecular dynamic (MD) simulations facilitate understanding of the effect of surface morphology on the adhesive properties of polymer films. Kisin et al. (2007) made a molecular dynamics study to estimate the work of adhesion at a copper metal and (acrylonitrile-butadiene styrene) (ABS) interface. They found that the highest work of adhesion was between an oxidized copper surface and a high oxygen content copolymer, poly (styrene-alt-maleic anhydride) (SMAh). The simulation predicted that the work of adhesion for this interface is about 3.36 J/m^2. Using MD simulation Miura et al. (2019) have investigated the effect of nanoporous structures on mechanical responses at polymer/metal interfaces. The results predicted that porous substrates would exhibit higher shear forces than flat substrates. They also reported that the interface area, the pulling mode, pore size, and polymer chain length play significant roles in the adhesion forces occuring between polymers and rough metal surfaces. Alberto et al. (Baggioli, 2020) have simulated the polymer-mediated adhesion of nanoscale surface morphology and failure mechanisms. The results suggested that the cohesive behaviour of model adhesive films relies on polymer/surface interactions and the morphology of the adherend at the nanoscale level. They also reported that a small variation in surface morphology could significantly affect the barrier for interfacial nucleation of nanoscale voids, leading to a change from a cohesive to a mixed adhesive–cohesive failure of the films. Ren et al. (Ren and Li, 2014) have simulated dynamic failure processes of bonded joint structures using a three-dimensional multiscale adhesive process zone model (APZM). They suggested that the APZM model is suitable for predicting the mechanical responses of adhesively bonded structures.This model also captures the failure behaviour of bonded joints at different dynamic loading conditions. They also validated the APZM model by testing a series of adhesively bonded joints under different loading conditions (shear, tension and mixed) to accurately predict the stresses in adhesive media.

5.8.9 Non-Destructive Test and Evaluation (NDT&E)

Adhesive bonded joining is a relatively special manufacturing process compared with conventional joining methods. Various defects may appear in bonded joints due to various process parameters. (Petrie, 2007; Özes and Nurhan Ne, 2015). Common defects in bonded joints are adhesive defects, cohesive defects, and bulk or gross defects. Adhesive defects are commonly due to poor adhesion between adherends and adhesive, voids, discontinuities and weak interfacial adhesion. These result in poor performance of the bonded joints. Cohesive defects depend greatly on the process parameters used in the fabrication of the adherends. Curing temperature, material storage condition, moisture and temperature degradation influence the formation of cohesive defects in bonded joints (Zhang et al. 2013).

In some cases, bulk defects like cracks and debonding at the interface also lower the bonded joints' performance. To produce high-performance bonding in structural units, quality assessment is essential. This can be achieved by applying proper inspection and measuring techniques to identify unbonded areas, delamination and correctly bonded joints. But conventional inspection techniques cannot identify concealed faults in bonded structures. These can be located using non-destructive test (NDT) measurements (Fuertes et al., 2015; Dillingham, 2015). NDT is widely used in the aerospace, oil production and exploration, chemical and automotive industries. Examples of NDT are standard ultrasonic immersion inspection (C-scan) and ultrasonic velocity measurement.

5.9 ENVIRONMENTAL DURABILITY OF ADHESIVELY BONDED JOINTS

Polymeric adhesives and polymer-based composites materials reach their limits under extreme environmental variations. Various in-service conditions influence the overall performance of the structural unit (Zhang et al., 2013; Broughton, 2012). Polymeric materials or structural adhesives become soft or rubbery at higher temperatures due to the high mobility of cross-linking networks. Differences in the thermal expansion coefficients of individual materials may cause the formation of stress concentration sites in interface regions, leading to lower performance. Prolonged exposure of polymeric materials to very low temperatures leads to excessive polymer embrittlement which results in the formation of microcracks in the polymer phase. Differential non-equilibrium thermal loading or temperature gradients such as thermal shock, thermal spike, freeze-thaw and thermal fatigue also severely influence mechanical performance (Ray and Rathore, 2014).

5.9.1 Effect of Temperature

In recent years, many researchers have focused on the development of high-performance structural adhesives for assembling multi-materials in various structural applications. Structural efficiency and durability depend on several factors, including temperature variation which severely influences structural integrity in real-world applications. Performance of the bonded structure mainly depends on adhesion behaviour and adhesive type, but polymeric adhesives alter their performance when exposed to elevated temperatures due to differences between the thermal expansion coefficient of the adhesive and substrate. All polymeric materials start degrading and become weaker at temperatures beyond the glass transition temperatures due to enhanced mobility of the polymeric chains (Prusty et al., 2015; Ray and Rathore, 2014). These adhesives lose strength when temperature increases and bonded joints show reduced stress transfer at higher temperatures due to the formation of microvoids, stress concentration and softening of the polymer.

5.9.2 Effect of Hydrothermal Ageing

Application of polymeric materials may be limited by the formation of bulk defects, surface cracks, and delamination caused by hydrothermal aging (Sánchez-Romate et al., 2020). When polymeric materials are aged in seawater, alkaline solution, acidic solution or distilled water degradation results from changes in the physical morphology of the network structure, plasticization, strength reduction effects and weakening of the interfacial region (Merdas et al., 2002). This may be attributed to hydrolysis and diffusion of water into the materials (molecules migrate from a higher concentration zone to a lower concentration zone). Prolonged exposure of polymeric and composite materials to hot and wet conditions leads to weakening and breakage of polymer crosslinking networks, swelling, cracking by osmosis and molecular degradation. Performance decreases and lifetime (durability) is reduced. Thermal gradients (sub-zero temperature to elevated temperature) damage the fibre/matrix interfacial region (Nayak, 2019; Toscano et al., 2016). When measuring the kinetics of water absorption into, and the saturation behaviour of polymeric materials, the percentage of water absorption can be calculated by using the following Eq. (5.2).

$$\text{Moisture uptake percentage} = M_t(\%) = \frac{M_t - M_O}{M_o} \times 100 \tag{5.2}$$

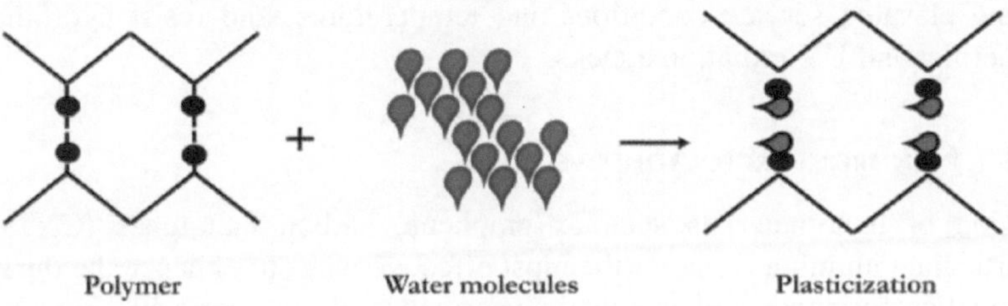

FIGURE 5.12 Schematic representation of plasticization behaviour of polymer (adopted from Hammiche, 2013) (Hammiche et al., 2013).

Where M_t is the mass of polymer sample after an aging time t and M_O is the mass of polymer specimen before an aging.

5.9.3 Effect of Hygrothermal Ageing

Moisture diffusion into polymeric materials rapidly influences their performance. Several factors including moisture, temperature, relative humidity, osmotic cracking, exposure time, material composition and matrix plasticization affect performance degradation during hygrothermal aging. Higher temperature increases network chain mobility and the moisture diffusion coefficient due to the coefficient of thermal expansion (CTE). Combined stress and aging conditions degrade polymeric materials more rapidly, impacting the performance of adhesively bonded joints. When a polymeric material is exposed to hygrothermal ageing the water molecules act as plasticizers (shown in Figure 5.12), which damages the polymer macromolecular chain. This causes poor mechanical performance of polymeric adhesive materials and lowers the glass transition temperature (Toscano et al., 2016; Hammiche et al., 2013).

In structural adhesives or polymeric materials exposed to hygrothermal aging, mechanical and thermal performance was severely affected by the swelling caused by absorbed moisture and solvent. Internal stresses developed during the aging process may be the origin of swelling action (Fang et al., 2018). The absorbed water tries to occupy the free space or free volume in the matrix material, which may alter the polymer links. This may be attributed to the formation of polar bonds due to chemical reactions between the polymeric network and free water molecules.

Figure 5.12 represents the plasticization behaviour of polymer. Combined stress and aging conditions degrade polymeric materials more rapidly, impacting the performance of adhesively bonded joints. When a polymeric material is exposed to hygrothermal ageing the water molecules act as plasticizers, which damages the polymer macromolecular chain and results in poor mechanical performance of polymeric adhesive materials.

5.10 NOVELTY AND ADVANCEMENTS IN POLYMERIC MATERIALS

New processing approaches to develop advanced polymeric adhesive materials include solution processing/matrix modification, fibre decoration by electrophoretic deposition and polymer blending. The aim is to create sustainable materials that can withstand elevated service conditions and temperatures, and resist hydrothermal, hygrothermal and UV radiation attacks.

5.10.1 Nanoreinforced Adhesive

Dispersion of nanomaterials such as graphene, carbon nanotubes (CNTs) and nanoparticulate alumina is one of the most effective ways to enhance the durability, mechanical performance and chemical performance of polymeric adhesives used in structural applications. The high aspect ratio and surface area of the nano-reinforcements enhances the overall performance of polymeric adhesives containing these additions in comparison with neat adhesives. Water absorption is also greatly reduced owing to chemical and physical interactions between the polymeric adhesive and nano reinforcement. Various researchers have reported that the addition of nano reinforcement material improves the mechanical response of polymeric adhesive. Nayak et al. (2019) reported that embedding of Al_2O_3 into thermoset epoxy polymer significantly improved its flexural performance, reduced water diffusivity, and enhanced hydrothermal resistance. Romate et al. (2020) have reported that addition of CNTs as a reinforcement material to structural adhesives decreases the degradation rate and reduces chain mobility in adhesive materials. The addition lowers the free volume space due to strong entanglements and improves the interface between adhesive and CNT reinforcement. Zielecki et al. (2017) reported that MWCNT-reinforced epoxy adhesive significantly improves the fatigue strength response and fatigue life due to improvement of the interfacial region and suggested that nanomaterial-reinforced epoxy adhesive is suitable for structural applications where high strength or high load-bearing capacity is required. Ganesh et al. (Ganesh Gupta K et al. 2020) reported embedding MWCNT in vinyl ester polymer to improve the flexural response of GFRP composites, and stated that smooth, effective stress transfer leads to improved performance due to the toughening and reinforcement effect of MWCNT. Ayatollahi et al. (2017) reported that the incorporation of reinforcement improves the shear strength and elongation at failure, with 0.2 wt.% CNT-reinforced epoxy and 0.8 wt.% nano-silica significantly improving the mechanical performance of bonded joints.

5.10.2 Interpenetrating Polymeric Network (IPN)

IPN consists of two or more polymers wherein the network formed is held by permanent entanglements. A thermoset-based epoxy and vinyl ester polymer blend significantly improved mechanical, thermal stability and damping characteristics. Ganesh et al. (Ganesh Gupta K et al. 2021) report that the GFRP composite's mechanical response is significantly enhanced through the IPN approach. They also suggest that this IPN process is both economical and a more prominent way of improving the adhesion of all synthetic and natural fibre material.

5.10.3 Adhesive Tapes

Adhesive tapes/adhesive films, thin layers of material used to assemble different components, are very suitable for automated manufacturing systems. The adhesive film approach, followed by liquid-type adhesives, lowers production costs and time compared with mechanical fastening. Polymeric adhesive materials have become in great demand for biomedical applications because of their versatile properties, such as lighter weight and biocompatibility. Biocompatible adhesives include fibrin glue (a combination of fibrinogen and thrombin), used successfully for sealing hemostasis (Spotnitz 2014). Spraying a thin layer of biocompatible material on the bleeding surface immediately controls the loss of blood. These surgical adhesives have some disadvantages, however, such as the risk of viral infection. Various studies have been undertaken in pursuit of the additional requirements of biomedical applications.

5.11 COMPLEX GEOMETRIES WHERE ADHESIVE CAN BE APPLICABLE COMPARED TO TRADITIONAL JOINTS

The adhesive bonding technique is widely recognized in various sectors as a more cost-effective technique than traditional joining methods due to the reduction in the overall structural weight of the component, improved fuel efficiency, and ease of assembly. Various complex structural applications include:

- *Aerospace*: Wing blades, fuselage body, tail surfaces, flaps, interior and exterior of plane, ailerons (B. Ganesh Gupta K et al. 2020).
- *Automotive*: Structural cabins, door assembly, glass panels, body and roof reinforcement panels, brake shoes, plastic bumpers, disk brake pads, radiator tanks, plastic load floors (Schneberger, 1990).
- *Medical*: Dental fillings and restorations
- *Electronic*: Wafer dicing, glop-top, snowboards, heat sinks, surface-mounted devices, plastic racing shells, chip encapsulation, potting and lens bonding (Connolly 2008)
- *Energy*: Composite blades assembly process
- *Oil and Gas*: Downhole sensor, gaskets, pumps and valves.

5.12 ADVANTAGES OF ADHESIVELY BONDED STRUCTURAL JOINTS OVER MECHANICAL FASTENING

The high demand for adhesive technology in various structural industries is due to its versatile performance. Adhesive joining is inexpensive compared to traditional joining techniques where large bonding areas are required (Budhea et al., 2017; Wahab, 2016; Akpinar et al., 2017). It reduces the need for additional manufacturing processes like machining, drilling and milling operations and enables large structural areas to be easily joined. Advantages of adhesive bonded joints include: simplified design process, cost reduction, ease of assembly, limited number of parts for assembly, smooth appearance, good damping behaviour, absence of galvanic corrosion, improved thermal insulation, better energy absorption, noise reduction, uniform stress distribution, high structural integrity, flexibility of material geometry and multi-material combinations.

Automation improves the production rate and can be achieved with the use of robotic arms to mix adhesive components and apply them more accurately to parts of structures where assembly requires polymeric adhesive. Adhesives are able to flexibly produce strong and stiff honeycomb structures, in which the honeycomb core adheres to the facing panel materials, and this adherend material may be either metallic, composite or both. Honeycomb applications include aerospace (gliders, rocket substructure, telescope mirror structure) and sports (snowboards, racing shells).

Traditional mechanical joining methods such as rivets, pins, nuts and bolts, etc., have more limited applications, especially in both thin sections and complex structures. Disadvantages include:

- Additional weight
- Possible galvanic corrosion due to dissimilarity in material composition
- Low sealing in complicated structural designs.
- Drilling operations required, which may lead to the beginnings of stress concentration and subsequent damage to the structural integrity.
- Holes required for mechanical two-piece bolt assembly cause delamination of composite material which may lead to internal stresses or stress concentration, resulting in fatigue cracks and lower structural efficiency.

5.13 CONCLUSION

This chapter has examined the principles of adhesively bonded structural joints and the factors affecting their mechanical and thermal responses. Different types of adherend surface treatments and their use in emerging areas have been described, together with design methods, the effect of water and temperature on durability, and the major failure strength tests that are used to determine intrinsic adhesive properties.

ACKNOWLEDGEMENTS

The authors expressed their heartfelt gratitude to the National Institute of Technology Rourkela, India, for providing the necessary facilities during this research work.

REFERENCES

Akpinar, I. A., K. Gültekin, S. Akpinar, H. Akbulut, and A. Ozel. 2017. Research on strength of nanocomposite adhesively bonded composite joints. *Composites Part B: Engineering* 126: 143–152. doi:10.1016/j.compositesb.2017.06.016.

Arenas, J. M., C. Alía, J. J. Narbón, R. Ocaña, and C. González. 2013. Considerations for the industrial application of structural adhesive joints in the aluminium-composite material bonding. *Composites Part B: Engineering* 44 (1): 417–423. doi:10.1016/j.compositesb.2012.04.026.

Ayatollahi, M. R., A. Nemati Giv, S. M. J. Razavi, and H. Khoramishad. 2017. Mechanical properties of adhesively single lap-bonded joints reinforced with multi-walled carbon nanotubes and silica nanoparticles. *Journal of Adhesion*. 93. Doi:10.1080/00218464.2016.1187069.

Baggioli, A. 2020. *Polymer-Mediated Adhesion: Nanoscale Surface Morphology and Failure Mechanisms*. Washington, DC: ACS Publications.

Banea, M.D. and L.F.M. Da Silva. 2009. Adhesively bonded joints in composite materials: an overview. *Proceedings of the Institution of Mechanical Engineers, Part L: Journal of Materials: Design and Applications* 223 (1): 1–18. Doi:10.1243/14644207JMDA219.

Banea, M. D., L. F.M. Da Silva, R. Carbas, and R. D.S.G. Campilho. 2017. ffect of material on the mechanical behaviour of adhesive joints for the automotive industry. *Journal of Adhesion Science and Technology* 31 (6): 663–676. Doi:10.1080/01694243.2016.1229842.

Broughton, W. 2012. Testing the mechanical, thermal and chemical properties of adhesives for marine environments. *Adhesives in Marine Engineering*. Doi:10.1533/9780857096159.2.99.

Budhea, S., M.D. Baneaa, S. de Barrosa, and L.F.M. da Silvab. 2017. An updated review of adhesively bonded joints in composite materials. *International Journal of Adhesion and Adhesives* 72: 30–42. Doi:10.1016/j.ijadhadh.2011.02.003.

Connolly, C. 2008. Adhesives in electronic and electrical assembly. *Assembly Automation* 28 (4): 289–294. Doi:10.1108/01445150810904431.

Dillingham, R. G. 2015. *Composite bond inspection. Structural Integrity and Durability of Advanced Composites: Innovative Modelling Methods and Intelligent Design*. Elsevier Ltd. Doi:10.1016/B978-0-08-100137-0.00025-0.

Ebnesajjad, S. 2009. *Adhesives Technology Handbook*. Doi:10.1115/1.3225943.

Encinas, N., B. R. Oakley, M. A. Belcher, K. Y. Blohowiak, R. G. Dillingham, J. Abenojar, and M. A. Martínez. 2014. Surface modification of aircraft used composites for adhesive bonding. *International Journal of Adhesion and Adhesives* 50: 157–163. Doi:10.1016/j.ijadhadh.2014.01.004.

Fang, Y., P. Chen, R. Huo, Y. Liang, L. Wang, and W. Liu. 2018. Hygrothermal ageing of polymeric sandwich structures used in structural engineering. *Construction and Building Materials* 165: 812–824. Doi:10.1016/j.conbuildmat.2018.01.072.

Ganesh Gupta K, B. N. V. S., M. M. Hiremath, R. K. Prusty, and B. C. Ray. 2020a. Development of advanced fiber-reinforced polymer composites by polymer hybridization technique: emphasis on cure kinetics, mechanical, and thermomechanical performance. *Journal of Applied Polymer Science*. Doi:10.1002/app.49318.

Ganesh Gupta K, B. N. V. S., M. M. Hiremath, B. C. Ray, and R. K. Prusty. 2020b. Improved mechanical responses of GFRP composites with epoxy-vinyl ester interpenetrating polymer network. *Polymer Testing*, 107008. Doi:10.1016/j.polymertesting.2020.107008.

Ganesh Gupta K, B. N. V. S., M. M. Hiremath, A. O. Fulmali, R. K. Prusty, and B. C. Ray. 2020c. Investigation of adhesively bonded multi-material joints: an assessment on joint efficiency and fracture morphology. *Materials Today: Proceedings*. doi:10.1016/j.matpr.2020.02.074.

Ganesh Gupta K, B. N. V. S., M. M. Hiremath, B. Sen, R. K. Prusty, and B. C. Ray. 2020d. Influence of loading rate on adhesively bonded tin-glass/epoxy single lap joint. *Materials Today: Proceedings*, 26: 1850–1854. Doi:10.1016/j.matpr.2020.02.406.

Ganesh Gupta K, B. N. V. S., A. Yadav, M. M. Hiremath, R. K. Prusty, and B. C. Ray. 2020e. Enhancement of mechanical properties of glass fiber reinforced vinyl ester composites by embedding multi-walled carbon nanotubes through solution processing technique. *Materials Today: Proceedings* 27: 1045–1050. Doi:10.1016/j.matpr.2020.01.391.

Ganesh Gupta K, B. N. V. S., M. M. Hiremath, A. O. Fulmali, R. K. Prusty, and B. C. Ray. 2020f. Multimaterial laminated composites: an assessment of effect of stacking sequence on flexural response. *Materials Today: Proceedings*. Doi:10.1016/j.matpr.2020.08.547.

Hammiche, D., A. Boukerrou, H. Djidjelli, Y. M. Corre, Y. Grohens, and I. Pillin. 2013. Hydrothermal ageing of alfa fiber reinforced polyvinylchloride composites. *Construction and Building Materials* 47: 293–300. Doi:10.1016/j.conbuildmat.2013.05.078.

Jeevi, G., S. K. Nayak, and M. A. Kader. 2019. Review on adhesive joints and their application in hybrid composite structures. *Journal of Adhesion Science and Technology* 33(14): 1497–1520. doi:10.1080/01694243.2018.1543528.

Jensen, R., D. Deschepper, D. Flanagan, G. Chaney, and C. Pergantis. 2016. *Adhesives: Test Method, Group Assignment, and Categorization Guide for High-Loading-Rate Applications – Preparation and Testing of Single Lap Joints (Ver. 2.2, Unlimited)*, 1–66.

Kanerva, M. and O. Saarela. 2013. The peel ply surface treatment for adhesive bonding of composites: a review. *International Journal of Adhesion and Adhesives* 43: 60–69. Doi:10.1016/j.ijadhadh.2013.01.014.

Kisin, S., J. B. Vukić, P. G. T. Van Der Varst, G. De With, and C. E. Koning. 2007. Estimating the polymer-metal work of adhesion from molecular dynamics simulations. *Chemistry of Materials* 19 (4): 903–907. doi:10.1021/cm0621702.

Küçükoğlu, A. and F. Karpat. 2017. *The joining techniques for thermoplastics materials. Proceedings of the ASME 2016 International Mechanical Engineering Congress and Exposition* 11–17: 1–7.

Lathabai, S. 2011. Joining of aluminium and its alloys. *Woodhead Publishing Series in Metals and Surface Engineering. Woodhead Publishing Limited.* Doi:10.2464/jilm.1956.19_80.

LeBacq, C., Y. Brechet, H. R. Shercliff, T. Jeggy, and L. Salvo. 2002. Selection of joining methods in mechanical design. *Materials and Design* 23 (4): 405–416. doi:10.1016/s0261-3069(01)00093-0.

Li, H., H. Liang, F. He, Y. Huang, and Y. Wan. 2009. Air dielectric barrier discharges plasma surface treatment of three-dimensional braided carbon fiber reinforced epoxy composites. *Surface and Coatings Technology* 203 (10–11): 1317–1321. doi:10.1016/j.surfcoat.2008.10.042.

Merdas, I., F. Thominette, A. Tcharkhtchi, and J. Verdu. 2002. Factors governing water absorption by composite matrices. *Composites Science and Technology* 62 (4): 487–492. Doi:10.1016/S0266-3538(01)00138-5.

Messler, R. W. 2004. Mechanical fasteners, integral attachments, and other mechanical joining methods. *Joining of Materials and Structures*, 105–176. doi:10.1016/b978-075067757-8/50003-8.

Miura, T., M. Funada, Y. Shimoi, and H. Morita. 2019. Simulation study of the effects of nanoporous structures on mechanical properties at polymer-metal interfaces. *Journal of Physical Chemistry B* 123 (5): 1161–1170. Doi:10.1021/acs.jpcb.8b10556.

Nayak, R. K. 2019. Influence of seawater aging on mechanical properties of nano-Al_2O_3 embedded glass fiber reinforced polymer nanocomposites. *Construction and Building Materials* 221: 12–19. Doi:10.1016/j.conbuildmat.2019.06.043.

Oldewurtel, A. 2019. Surface treatment methods. *Mold-Making Handb for the Plast Eng*, 351–372. Doi:10.1016/B978-0-12-817010-6.00003-5.

Özes, Ç. and Nurhan Ne, G. 2015. *Experimental Study on Steel to FRP Bonded Lap* 2015.

Petrie, Edward M. 2007. An introduction to adhesive and sealants. *Handbook of Adhesives and Sealants*, 1–422. www.digitalengineeringlibrary.com.

Prusty, R. K., D. K. Rathore, and B. C. Ray. 2017. CNT/Polymer interface in polymeric composites and its sensitivity study at different environments. *Advances in Colloid and Interface Science* 240: 77–106. doi:10.1016/j.cis.2016.12.008.

Prusty, R. K., D. K. Rathore, M. J. Shukla, and B. C. Ray. 2015. Flexural behaviour of CNT-filled glass/epoxy composites in an in-situ environment emphasizing temperature variation. *Composites Part B: Engineering* 83: 166–174. doi:10.1016/j.compositesb.2015.08.035.

Ray, B. C. and D. Rathore. 2014. Durability and integrity studies of environmentally conditioned interfaces in fibrous polymeric composites: critical concepts and comments. *Advances in Colloid and Interface Science* 209: 268–283. doi:10.1016/j.cis.2013.12.014.

Reis, P. N. B., J. A. M. Ferreira, and F. Antunes. 2011. Effect of adherends rigidity on the shear strength of single lap adhesive joints. *International Journal of Adhesion and Adhesives* 31 (4): 193–201. doi:10.1016/j.ijadhadh.2010.12.003.

Ren, B. and S. Li. 2014. Multiscale modeling and prediction of bonded joint failure by using an adhesive process zone model. *Theoretical and Applied Fracture Mechanics* 72 (1): 76–88. doi:10.1016/j.tafmec.2014.04.007.

Ribeiro, T. E. A., R. D.S. G. Campilho, L. F. M. da Silva, and L. Goglio. 2016. Damage analysis of composite-aluminium adhesively-bonded single-lap joints. *Composite Structures* 136: 25–33. Doi:10.1016/j.compstruct.2015.09.054.

Saeedikhani, M., M. Javidi, and A. Yazdani. 2013. Anodizing of 2024-T3 aluminum alloy in sulfuric-boric-phosphoric acids and its corrosion behavior. *Transactions of Nonferrous Metals Society of China (English Edition)* 23 (9): 2551–2559. Doi:10.1016/S1003-6326(13)62767-3.

Saleema, N., D. K. Sarkar, R. W. Paynter, D. Gallant, and M. Eskandarian. 2012. A simple surface treatment and characterization of AA 6061 aluminum alloy surface for adhesive bonding applications. *Applied Surface Science* 261: 742–748. doi:10.1016/j.apsusc.2012.08.091.

Sánchez-Romate, X. F., P. Terán, S. G. Prolongo, M. Sánchez, and A. Ureña. 2020. Hydrothermal ageing on self-sensing bonded joints with novel carbon nanomaterial reinforced adhesive films. *Polymer Degradation and Stability* 177. doi:10.1016/j.polymdegradstab.2020.109170.

Fuertes, S., A. Theodor, T. Kruse, T. Körwien, and M. Geistbeck. 2015. Bonding of CFRP primary aerospace structures - discussion of the certification boundary conditions and related technology fields addressing the needs for development. *Composite Interfaces* 22 (8): 795–808. doi:10.1080/09276440.2015.1077048.

Schneberger, G. L. 1990. Adhesives in the automobile industry. *Handbook of Adhesives*, 729–735. doi:10.1007/978-1-4613-0671-9_45.

da Silva, Lucas F.M., R. J. C. Carbas, G. W. Critchlow, M. A.V. Figueiredo, and K. Brown. 2009. Effect of material, geometry, surface treatment and environment on the shear strength of single lap joints. *International Journal of Adhesion and Adhesives* 29 (6): 621–632. doi:10.1016/j.ijadhadh.2009.02.012.

Spotnitz, W. D. 2014. Fibrin sealant: the only approved hemostat, sealant, and adhesive—a laboratory and clinical perspective. *ISRN Surgery* 2014: 1–28. Doi:10.1155/2014/203943.

Streitferdt, A., N. Rudolph, and I. Taha. 2017. Co-curing of CFRP-steel hybrid joints using the vacuum assisted resin infusion process. *Applied Composite Materials* 24 (5): 1137–1149. Doi:10.1007/s10443-016-9575-3.

Toscano, A., G. Pitarresi, M. Scafidi, M. Di Filippo, G. Spadaro, and S. Alessi. 2016. Water diffusion and swelling stresses in highly crosslinked epoxy matrices. *Polymer Degradation and Stability* 133: 255–263. Doi:10.1016/j.polymdegradstab.2016.09.004.

Troughton, M. J. 2008. Handbook of plastics and joining (mechanical fastening). *Handbook of Plastics Joining*: 175–201. doi:10.1016/b978-0-8155-1581-4.50020-2.

Wahab, M. A. 2016. *JOINING COMPOSITES with ADHESIVES Theory and Applications*, 1–316.

Wu, X., K. He, Z. Gong, Z. Liu, and J. Jiang. 2020. The shear strength of composite secondary bonded single-lap joints with different fabrication methods. *Journal of Adhesion Science and Technology* 34 (9): 936–948. doi:10.1080/01694243.2019.1690775.

Zhang, Fan, X. Yang, H. P. Wang, X. Zhang, Y. Xia, and Q. Zhou. 2013. Durability of adhesively-bonded single lap-shear joints in accelerated hygrothermal exposure for automotive applications. *International Journal of Adhesion and Adhesives* 44: 130–137. doi:10.1016/j.ijadhadh.2013.02.009.

Zielecki, W., A. Kubit, T. Trzepieciński, U. Narkiewicz, and Z. Czech. 2017. Impact of multi-wall carbon nanotubes on the fatigue strength of adhesive joints. *International Journal of Adhesion and Adhesives* 73 (November 2016): 16–21. doi:10.1016/j.ijadhadh.2016.11.005.

6 Damage Identification of Natural Fibre Composites Using Modal Parameters

D. Mallikarjuna Reddy
Vellore Institute of Technology, Vellore, India

G. Venkatachalam
Vellore Institute of Technology, Chennai, India

Shivasharanayya Swamy
REVA University, Bangalore, India

J Naveen
Vellore Institute of Technology, Vellore, India

Jose Machado
University of Minho, Portugal

CONTENTS

DOI: 10.1201/9781003128861-6

6.1 INTRODUCTION

Natural fibres are derived from the stem, leaf and root of plants (Manimaran et al., 2018), and their advantages include availability, light weight and biodegradability. Natural fibres such as hemp, jute and flax are used as reinforcement in polymer composites (Saravanan et al., 2016) and also as replacements for glass fibres in many industries including the automotive sector (Habibi et al., 2019a). Vehicle components that are subjected to impact benefit from the stiffness of the material. Hemp fibre is rich in cellulose compared to other natural fibres and performs with high strength. Hemp fibres are widely used in such applications as car door, floor and body panels (Liu et al., 2017; Baghaei et al., 2014). Chaudhary (2018) fabricated a hybrid composite with a combination of hemp, jute and flax natural fibre and concluded that natural fibres exhibit good performance in terms of tensile, impact and flexural strength compared to conventional glass fibres. Shahzad (2012) concluded that hemp fibre has more strength than synthetic fibre.

Impact causes fibre damage, matrix cracking and permanent indentation (Sun and Hallett, 2018). Patil et al. (2018) studied the effect of low-velocity impact on composite materials. Davies and Olsson (2004) and Habibi et al. (2019b) conducted low-velocity impact tests by varying different parameters and observing the different types of damage in the specimens. Rakesh et al. (2012) conducted an experimental study on bi-directional woven glass fibres and found the optimal results for improved impact resistance. Ahmed and Wei (2015) conducted research on the dynamic and static response of composites structures subjected to low-velocity impact and quasi-static loads. Habibi et al. (2019b) determined the effect of impact damage on woven flax fibre, based on material deformation, tensile properties and local strain evolution. Patil and Reddy (2020c) identified damage in the impacted composite laminates using the vibration-based damage index method. Patil and Reddy (2020a) analyzed both normal and oblique impacts on composite plates.

Many analytical methods are used in the literature to identify damage to structures. Lele and Maiti (2002) studied crack detection in a short beam using the vibration method. Many researchers prefer using finite element methods to identify structural damage (Chinchalkar, 2001; Brighton, 2011; Castel et al., 2012; Dona et al., 2015). Many experiments have sought to verify the analytical or numerical model (Sinha et al., 2002; Baghiee et al., 2009). Pastor et al. (2012) presented the correlation between the natural frequency and modal shapes. Ewins (2001) compared experimental and numerical values, where MAC is used for the numerical analysis.

In the study reported here, natural composite laminates were fabricated using the hand lay-up method with considered epoxy as a matrix. After fabricating, impact tests were carried out on hemp-reinforced epoxy composite, and modal analysis was performed to identify damage to the specimen.

6.2 MATERIALS AND METHODS

6.2.1 Hemp Fibre (*Cannabis sativa*)

Hemp is a natural fibre plant, like jute, sisal and banana. It is a tall-growing variety of the cannabis plant that grows to between 4 and 15 ft in height. The hemp plant is surrounded by core fibre. Hemp fibre consists of cannabinoids, tetrahydrocannabinol

FIGURE 6.1 (a) Hemp plant; (b) Hemp woven fabric mat.

TABLE 6.1
Chemical Properties of Hemp Fibre

Chemical properties	Value
Cellulose	70–78
Biomass	8–32
Moisture	7–14
Pectin	2–20
Lignin	2–10
Hemicellulose	8–20

TABLE 6.2
Mechanical and Physical Properties of Hemp Fibre

Properties	Value
Diameter of the fibre (μm)	300–800
Length of the fibre (mm)	9–15
Young's modulus (GPa)	35–70
Density (g/cm^3)	1.4–1.6
Tensile strength (MPa)	200–1000
Tensile modulus	24–90
Elongation in %	1–3.5
Failure strain	2–5

and cannabidiol. Figure 6.1 shows the hemp plant and hemp mat; mechanical and chemical properties are given in Tables 6.1 and 6.2.

6.2.2 Specimen Preparation

The hemp and jute fibres were reinforced with epoxy resin (LY556) and hardener (HY951). The natural fibre to matrix ratio was 1:1 and the epoxy resin to hardener

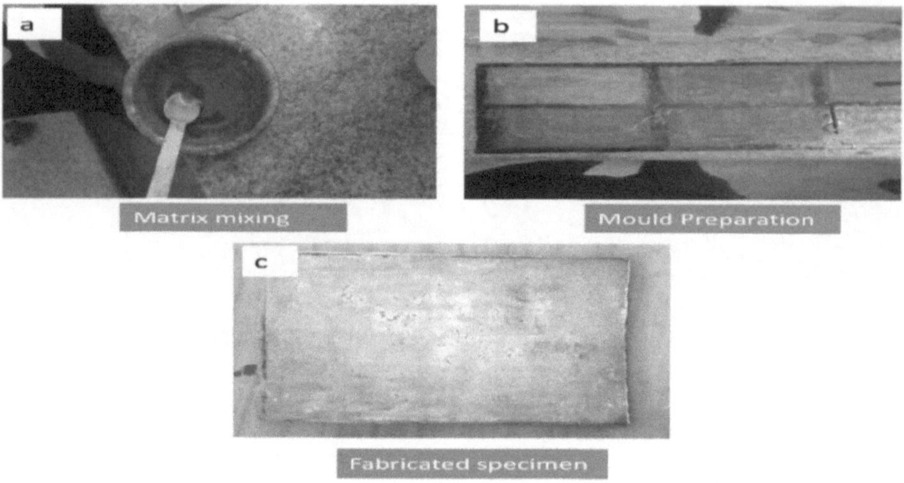

FIGURE 6.2 Specimen preparation (a) Matrix mixing; (b) Mould preparation; (c) Fabricated hemp laminate.

TABLE 6.3
Specification of Different Specimens

Sl. No.	Specimen ID	Specimen type
1	SH1	Specimen without damage
2	SD1(Centre)	Damage at the centre
3	SD2 (Edge)	Damage at the edge
4	SD3 (Multi-damage)	Damage in 3 layers

ratio was maintained at 10:1 by weight with the addition of filler material. Laminates were prepared using the hand lay-up method. The fabricated plate measures 150 mm × 150 mm and is 3 mm thick (Figure 6.2). Four different test specimens were taken for modal analysis (for specimen specifications, see Table 6.3). For the modal analysis, a total of 36 response points, including natural frequency and mode shape, were taken into account to obtain modal parameters (Figure 6.3).

6.2.3 Low-velocity Impact (LVI)

The low-velocity impact test is carried out in accordance with standard ASTM D7136 using drop weight impact (Figure 6.4). A conical shape impactor with mass of 3.5 kg is used.

6.2.4 Modal Analysis

Modal analysis is performed on the fabricated composite specimens using an impact hammer. The modal test set-up is shown in Figure 6.5. The main aim in modal testing

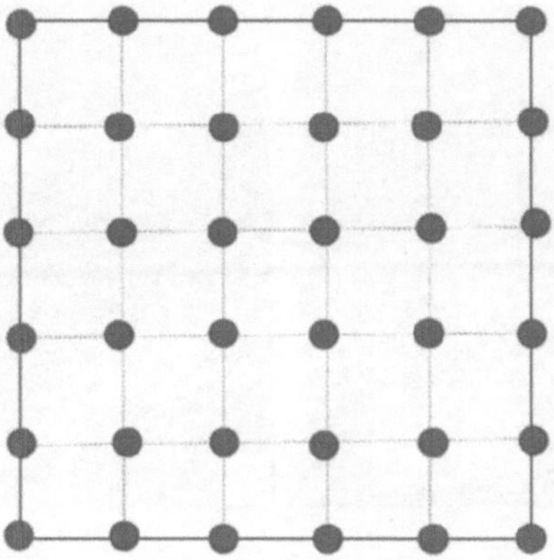

FIGURE 6.3 Response points considered for modal analysis.

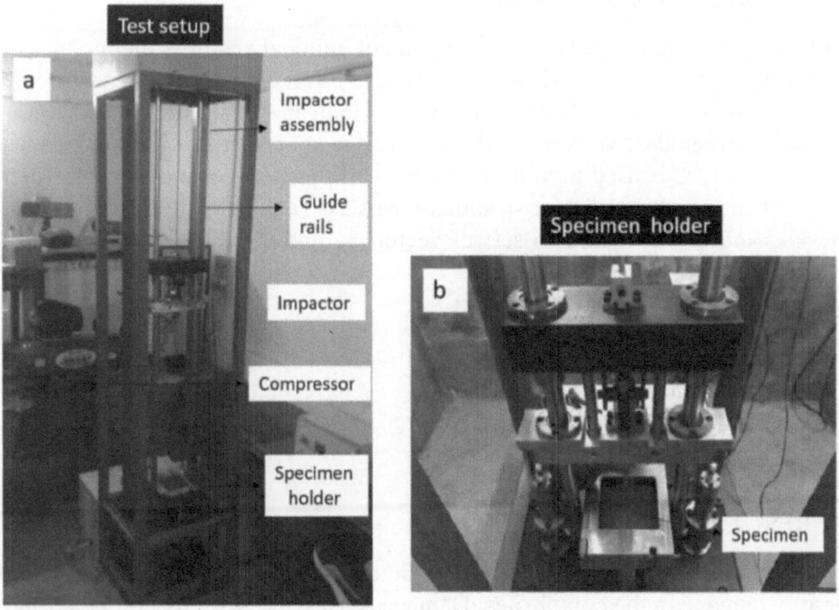

FIGURE 6.4 Drop weight impact test set-up.

is to establish the modal parameters, such as natural frequency of the specimen, through the mechanical wave. The accelerometer and hammer are connected to the data acquisition system and collect frequency response function (FRF) signals. In modal testing, the accelerometer is placed at the centre of the plate and exits through the impact hammer. The signals received from the system are analyzed and displayed as Fast Fourier Transform (FFT) data using Dewesoft software.

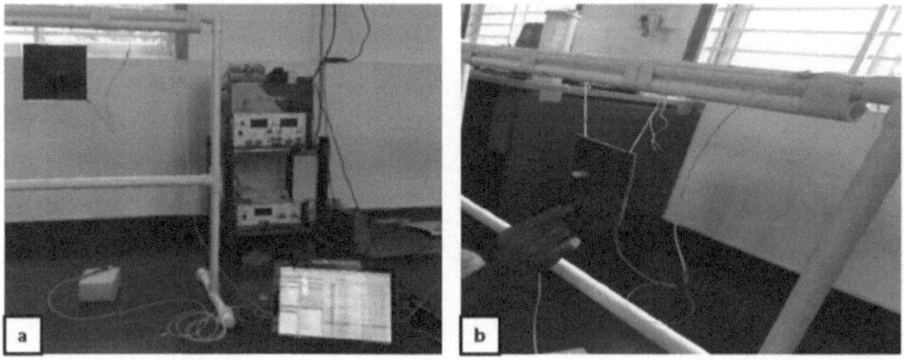

FIGURE 6.5 (a) Experimental modal analysis set-up; (b) Close-up view of experimental modal analysis set-up.

6.2.5 Modal Assurance Criterion

One of the most popular tools for comparison of a modal vector is the modal assurance criterion (MAC). It is used as a mode shape correlation to quantify the accuracy of the paired modal vectors. MAC is a statistical indicator of sensitivity to the largest difference between the two compared values and is not sensitive to small changes in the structure. MAC is used to find the difference in the exit location by considering the mode shapes of the particular specimens. MAC is calculated from Equation (6.1), where ψ_{cqr} and ψ^{*}_{dqr} are the two sets of vectors to find the difference between two specimens.

$$MAC\left(cqr\right)=\frac{\left|\sum_{q=1}^{N}\psi_{cqr}\psi^{*}_{dqr}\right|^{2}}{\sum_{q=1}^{N}\psi_{cqr}\psi^{*}_{cqr}\sum_{q=1}^{N}\psi_{dqr}\psi^{*}_{dqr}} \tag{6.1}$$

6.3 RESULTS AND DISCUSSION

In this research, modal parameters were identified using vibration methods in order to identify damage in the composites. Damage sensitivity is discussed in the following section.

6.3.1 Modal Parameter Identification

Modal analysis is carried out to identify the modal parameters of the specimens. Using modal analysis, frequency response functions (FRFs) are collected that are sensitive to small changes and damage to the structure. First, 36 response points are obtained and stored using signal process Dewesoft software. It is very important to highlight the response points considered for vibration test, as illustrated in

TABLE 6.4
Experimental Natural Frequency for All Specimens

Natural frequency no	Natural frequency for SH1 (in Hz)	Natural frequency for SD1 (in Hz)	Natural frequency for SD2 (in Hz)	Natural frequency for SD3 (in Hz)
1	179	186	183	196
2	287	393	396	315
3	368	494	515	399
4	469	890	610	519
5	863	948	937	632
6	1460	1008	1020	953
7	1570	1160	1570	1210
8	1780	1510	1570	1580
9	1870	1820	1580	1610

Figure 6.3, which shows the impact hammer with the addition of an accelerometer installed close to the centre of the plate. When the reciprocity principle was applied to the received FRF, the natural frequency and mode shape were obtained (see Table 6.4). The modal analysis is performed with free-free boundary condition. Figure 6.6 represents the time history for specimen responses recorded from accelerometers. Natural frequency and mode shapes are obtained from modal analysis

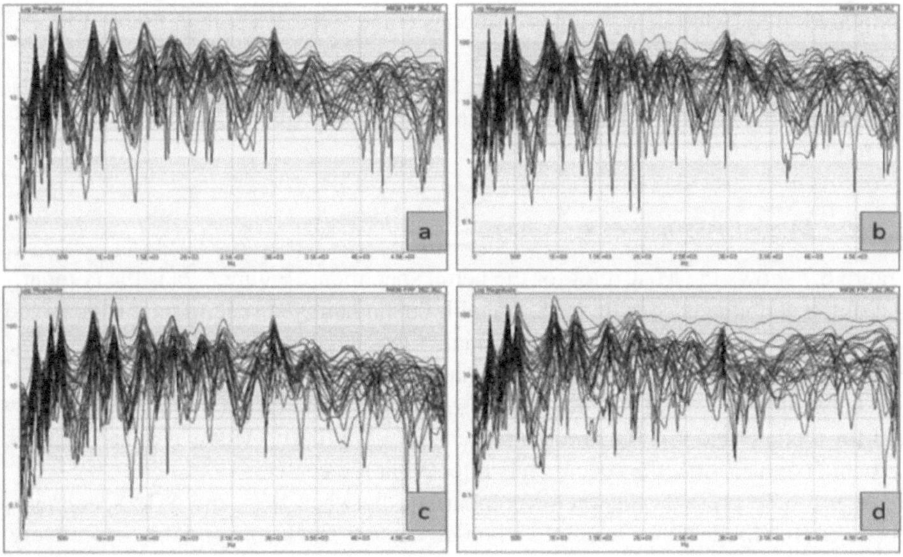

FIGURE 6.6 Experimentally obtained frequency response plot for all damage scenarios (a) SH1; (b) SD1; (c) SD2; (d) SD3.

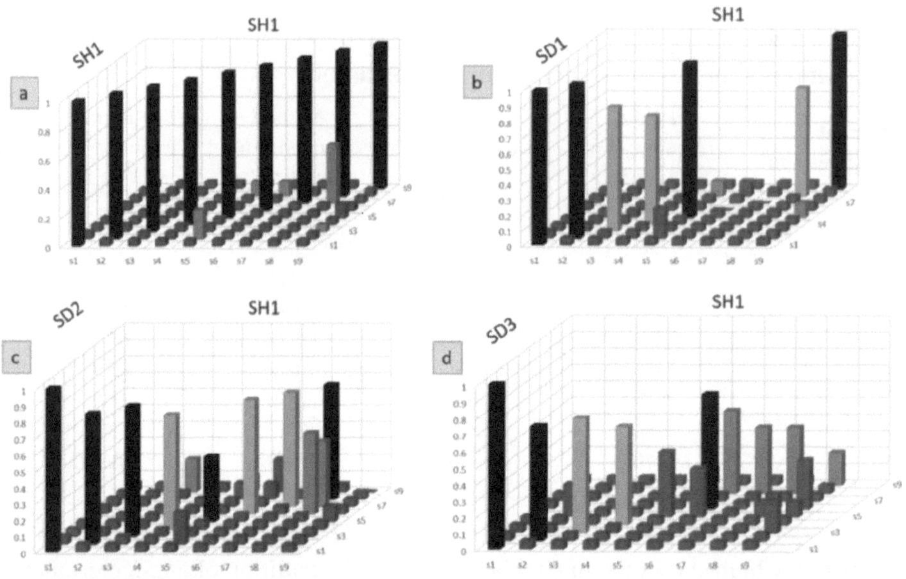

FIGURE 6.7 MAC plot for all specimens (a) SH1 (b) SD1 (c) SD2 (d) SD3 against SH1.

for all specimens. Modal parameters are also obtained using Dewesoft software (see Table 6.4).

Figure 6.6(a)–(d) shows the frequency responses of all specimens obtained through the accelerometer sensor. SD1shows that mode 1 and mode 5 are increased compared to SH1;frequency values of modes 6 and 9 are lower than those of modes 1 and 5.

FRF results for specimens SD1 and SD2 are shown in Figures 6.6b and 6.6c. The SH1 specimen shows an increase in the modal frequency till mode 5, then a decreases for modes 7–9.

6.3.2 Modal Assurance Criterion

Figure 6.7 shows the MAC plot for the hemp specimens. Figure 6.7a is the comparison of the undamaged specimen (SH1) with the undamaged specimen, it is observed that the MAC shows unity along diagonal. Also, Eigenvectors are in red and it is confirming that the model is consistent. Later, Figure 6.7b is the comparison of vectors between SH1 and SD1 specimen, observed that at modes of 1,2,5,8 shows unity and for mode 3 and 4 is 0.8 and 0.6 respectively. Similarly, Mode at 6 and 7 shows zero and confirms that the specimen is affected at the center.

Figure 6.7c presents the vector comparison between SH1 and SD2 specimens, observed that mode 1 shows unity and mode 2 and 3 shows 0.7 and 0.8 respectively. Mode 10 shows zero and conforming that edge failure for in the specimen. Figure 6.7d also observed that there is lesscorrelation between the two vectors at the center modes. Conclude that MAC plots are sensitive to the damage along the diagonals and give the information about damage present in the specimens.

6.4 CONCLUSION

Hemp (*Cannabis sativa*)-reinforced composite laminates are successfully fabricated using the hand lay-up method. In this study, the effect of impact damage and the modal analysis for the composite laminates were investigated experimentally and the presence of damage in the specimens predicted. The following conclusion can be drawn.

- It is observed that impact damage strongly affects the modal parameters of the structure.
- MAC plots provide information about damage present in the structures.
- From the MAC plot, it can be observed that the MAC value exists only when damage was applied asymmetrically and no mode shape distortion is observed from the structure.

FUNDING

This research work is funded by the Science and Engineering Research Board (SERB), Government of India under grant ECR/2017/000512.

ACKNOWLEDGEMENTS

The authors would like to acknowledge the Department of Science and Technology (DST) for financial support under the grant (ECR/2017/000512).

REFERENCES

Ahmed, A. and L. Wei. 2015. The low-velocity impact damage resistance of the composite structures-a review. *Reviews on Advanced Materials Science*, 40(2).

Baghaei, B., M. Skrifvars, M. Salehi, T. Bashir, M. Rissanen, and P. Nousiainen. 2014. Novel aligned hemp fibre reinforcement for structural bio composites: Porosity, water absorption, mechanical performances and viscoelastic behaviour. *Composites. Part A, Applied Science and Manufacturing*, 61: 1–12. doi:10.1016/j.compositesa.2014.01.017.

Baghiee, N., R. M. Esfahani, and K. Moslem. 2009. Studies on damage and FRP strengthening of reinforced concrete beams by vibration monitoring. *Engineering Structure*, 31: 875–893.

Bensadoun, F., D. Depuydt, J. Baets, I. Verpoest, and A. W. Van Vuure. 2017. Low velocity impact properties of flax composites. *Composite Structures*, 176 (Supplement C): 933–944. doi:10.1016/j.compstruct.2017.05.005.

Brighton, D. A. 2011. *Finite Element Analysis of an Intentionally Damaged Prestressed Reinforced Concrete Beam Repaired with Carbon Fiber Reinforced Polymers*, The University of Toledo, Toledo, OH.

Castel, A., T. Vidal, and R. François. (2012). Finite-element modeling to calculate the overall stiffness of cracked reinforced concrete beams. *Journal of Structural Engineering*, 138(7), 889–898.

Chang, K. C. and C. W. Kim. 2016. Modal-parameter identification and vibration-based damage detection of a damaged steel truss bridge. *Engineering Structures*, 122 (September): 156–173. doi:10.1016/j.engstruct.2016.04.057.

Chaudhary, V., P. K. Bajpai, and S. Maheshwari. 2018. Studies on mechanical and morphological characterization of developed jute/hemp/flax reinforced hybrid composites for structural applications. *Journal of Natural Fibers*, 15 (1): 80–97. doi:10.1080/15440478.2017.1320260.

Chinchalkar, S. 2001. Detection of the crack location in beams using natural frequencies. *Journal of Sound and Vibration*, 247: 417–429.

Davies, G.A.O. and R. Olsson. 2004. Impact on composite structures. *The Aeronautical Journal*, 108 (1089): 541–563. doi:10.1017/S0001924000000385.

Dona, M., A. Palmeri, M. Lombardo, and A. Cicirello. 2015. An efficient two-node finite element formulation of multi-damaged beams including shear deformation and rotatory inertia. *Computers & Structures* 147: 96–106.

Ewins, D. J. 2001. *Modal Testing: Theory, Practice and Application*. Research Studies Press Ltd. Baldock, Hertfordshire, England.

Habibi, M., L. Laperrière, and H. M. Hassanabadi. 2019a. Replacing stitching and weaving in natural fiber reinforcement manufacturing, Part 1: mechanical behavior of unidirectional flax fiber composites. *Journal of Natural Fibers*, 16(7) 1064–1076.

Habibi, M., S. Selmi, L. Laperrière, H. Mahi, and S. Kelouwani. 2019b. Damage analysis of low-velocity impact of non-woven flax epoxy composites. *Journal of Natural Fibers*, 1–10. doi:10.1080/15440478.2019.1584076.

Lele, S.P. and S. K. Maiti. 2002. Modeling of transverse vibration of short beams for crack detection and measurement of crack extension. *Journal of Sound and Vibration*, 257: 559–583.

Liu, M., A. Thygesen, J. Summerscales, and A. S. Meyer. 2017. Targeted pre-treatment of hemp bast fibres for optimal performance in biocomposite materials: a review. *Industrial Crops and Products*, 108: 660–683. doi:10.1016/j.indcrop.2017.07.027.

Manimaran, P., M. R. Sanjay, P. Senthamaraikannan, M. Jawaid, S. S. Saravanakumar, and R. George. 2018. Synthesis and characterization of cellulosic fiber from red banana peduncle as reinforcement for potential applications. *Journal of Natural Fibers*, 16: 768–780. doi:10.1080/15440478.2018.1434851.

Pastor, M., M. Binda, and T. Harčarik. 2012. Modal assurance criterion. *Procedia Engineering*, 48: 543–548. doi:10.1016/j.proeng.2012.09.551.

Patil, S. and D.M. Reddy. 2020a. Study of oblique low velocity impact on composite plate. *Materials Today: Proceedings*, doi:10.1016/j.matpr.2020.03.125.

Patil, S. and D.M. Reddy. 2020b. Damage identification in hemp fiber (Cannabissativa) reinforced composite plates using MAC and COMAC correlation methods: experimental study, *Journal of Natural Fibers*. doi:10.1080/15440478.2020.1764449.

Patil, S. and D.M. Reddy. 2020c. Impact damage assessment in carbon fiber reinforced composite using vibration-based new damage index and ultrasonic C-scanning method. *Structures*, 28: 638–650.

Patil, S., D. M. Reddy and M. Reddy. (2018, April). *Low velocity impact analysis on composite structures–A review*. In *AIP Conference Proceedings* (Vol. 1943, No. 1, p. 020009). AIP Publishing LLC.

Saravanan, N., P. S. Sampath, T. A. Sukantha, and T. Natarajan. 2016. Extraction and characterization of new cellulose fiber from the Agrowaste of Lagenaria Siceraria (Bottle Guard) plant. *Journal of Advances in Chemistry*, 12: 4382–4388. doi:10.24297/jac.v12i9.3991.

Shahzad, A. 2012. Hemp fiber and its composites – a review. *Journal of Composite Materials*, 46 (8): 973–986. doi:10.1177/0021998311413623.

Sinha, J. K., M. I. Friswell, and S. Edwards. 2002. Simplified models for the location of cracks in beam structure using measured vibration data. *Journal of Sound and Vibration*, 251: 13–38.

Sun, X. C. and S. R. Hallett. 2018. Failure mechanisms and damage evolution of laminated composites under compression after impact (CAI): experimental and numerical study. *Composites Part A: Applied Science and Manufacturing*, 104: 41–59. doi:10.1016/j.compositesa.2017.10.026.

Wang, T., O. Celik, F. N. Catbas, and L. M. Zhang. A frequency and spatial domain decomposition method for operational strain modal analysis and its application. *Engineering Structures*, 114: 104–112. doi:10.1016/j.engstruct.2016.02.011.

7 An Overview of Adhesive Bonded Composite Joint Failure

Critical Comparison of Co-Curing, Co-Bonding and Secondary Bonding

K.M. Manjunatha Swamy and H. Manjunath
Siddaganga Institute of Technology, Tumkuru, Karnataka, India

N. Shanmugavadivu
RVS College of Engineering and Technology, Coimbatore, Tamil Nadu, India

Marappan Shanmugasundaram
Jazan University, Kingdom of Saudi Arabia

CONTENTS

DOI: 10.1201/9781003128861-7

7.1 INTRODUCTION

Composite materials are usually affected by mechanical loads and environmental conditions (Ronald et al. 2010). Adhesive bonding using a bonding agent is one of the best methods for joining two adjacent surfaces (Baldan. 2012). Adhesive bonding is widely used for applications including space exploration, microelectronic packaging and in the aerospace industries (Bowen, et al. 1989).

The greatest challenge with advanced composites in the aeronautical industries is structural repair, due to inherent dynamic damage. Generally three methods are used to connect composite surfaces: co-bonding, co-curing and secondary bonding. Secondary bonding, which is preferred to co-bonding as fewer curing cycles and parts are involved, is adopted for composite structure repairs (Hedges et al. 2004).

Kihara et al. (2003) conducted experiments to determine the shear strength of a double-joint configuration adhesive layer under impact stress. The effect of thermal exposure on the adhesive strength of low-carbon steel was studied by Lin et al. (2013). Vaidya et al. (2006) evaluated the failure of adhesive bonded joints due to transverse impact loading, and observed that transfer loading results in a larger concentration of peel tension compared to in-plane loading. Aktas and Polat (2010) developed a way to enhance the strength of single-lap composite joints, using fibre as a bolt. Kim et al. (2005) conducted an experimental study and numerical evaluation to determine the kinds of damage caused by the out-plane effect on the overlap region and to describe the mechanisms that cause the damage. Weitsman (1977) conducted an analytical study of the effects of moisture and temperature at the interface between the composite material adhesive. Similar results were found by other researchers (Higuchi et al. 2003). On the other hand, as adhesive thickness decreases, the maximum stress is found to increase. This chapter looks in detail at the major factors influencing bonded joints, the various types of composite joints and environmental parameters affecting them.

7.2 ADHESIVE BONDED JOINTS IN COMPOSITE MATERIALS

The use of adhesive-reinforced joints is of particular interest for lightweight designs; they are used in the aircraft industry to reduce weight.

Fibre-strengthened materials have been increasingly used in recent years. Regular mechanical bonding procedures have been shown to produce fibre damage and early breaks in composite structures (Teng et al. 2012). Table 7.1. summarizes the advantages and disadvantages of adhesive and mechanical bonding. Adhesive bonding is extremely attractive for fibre-fortified designs because of its different focal points.

TABLE 7.1
Advantages and Disadvantages of Mechanical and Adhesive Bonding

	Advantages	Disadvantages
Mechanical bonding	• No need for surface preparation • Ease of component disassembly and reassembly	• Stress concentration is only around the bond • Increases weight of the structure
Adhesive bonding	• High strength-to-weight ratio • Uniform stress distribution • Higher damage tolerance • Reduces the cost of manufacturing • Provides flexibility in design	• Special surface preparation and non-destructive test is necessary • Permanent assembly in bonding process • Increases manufacturing time during cure • Minimum resistance at high temperatures • Attention to temperature, moisture, environmental and safety issues are required during the bonding process

Source: Teng et al. (2012).

7.3 ADHESIVE BONDING OF COMPOSITES

The bonding approach chosen has a significant impact on joint strength, so the most appropriate technique should be chosen based on individual requirements. The three adhesive bonding methods are secondary boning, co-curing and co-bonding (Kruse et al. 2015) (Figure 7.1).

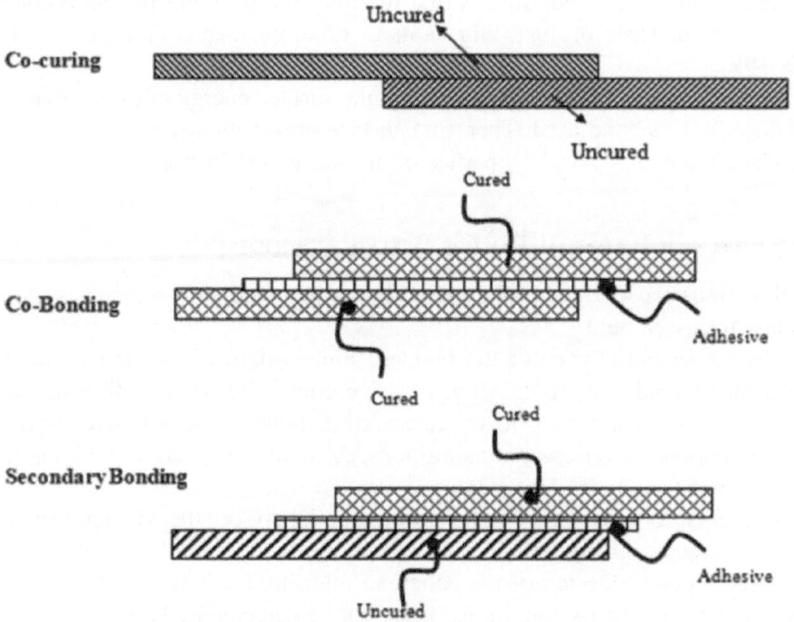

FIGURE 7.1 Adhesive bonding methods.

(i) *Co-curing* is the first step towards a fully integrated component. Chemical cross-linking is the joining method.
(ii) *Co-bonding* is the in-between stage of integration. The component is usually attached to one or more cured component along with an adhesive coating. Adhesion is the mechanism that connects the adhesive and the cured/uncured component. Chemical cross-linking occurs between the uncured component and the adhesive.
(iii) *Secondary bonding* is the latest method of integration. A film or paste adhesive is used to unite the two previously cured components. Adhesion is the linking mechanism between the adhesive and the adherend.

7.4 VARIABLES INFLUENCING THE REINFORCED JOINTS

The strength loss of a joint exposed to ambient temperature is primarily determined by moisture and heat. Standard tests show that adhesive joints lose their strength due to exposure to high temperatures and moisture (Kinloch 1983).

7.4.1 Effect of Surface Preparation in Reinforced Joints

Surface characteristics such as conductive physical and chemical conditions are critical to create strong interfacial bonds. Sample preparation is very important for bonded composite repairs and secondary bonding (Theodor et al. 2015). Surface preparation is usually required for composite materials before secondary bonding to improve interface adhesion. A common misconception regarding surface preparation prior to bonding is that the only requirement for good adhesion is a clean surface (free of dust, lubricants or other surface contaminants) (Frank et al. 2019).

In order to ensure quality surface wetting, the surface energy must be greater than that of the adhesive bond used. Therefore surface preparation is required for changing the chemical and physical properties of the surfaces (Claire et al. 2006).

7.4.2 Effect of Potential Failure Initiation Modes

Alhough various causes of early failure mode are observed, only the most significant effects are discussed here.

Disbond is an initial area of the bonded joints where connection between the adherend and the adhesive is lacking. Massive contamination of adherend surface and failures are the common primary causes of disbond in the adhesive application process. A disbond is observable using non-destructive inspection methods (NDI) such as ultrasonic study (Figure 7.2).

Weak bond is a bonded joint that has low strength between the adherend and adhesive. It is characterised by an adhesive failure mode. The fundamental reason for weak bonds is poor adhesion of the adherend interfaces as a result of surface contamination or poor process conditions. It cannot be detected by NDI due to the lack of a detectable interface layer (Shanmugavadivu 2019) (Figure 7.3).

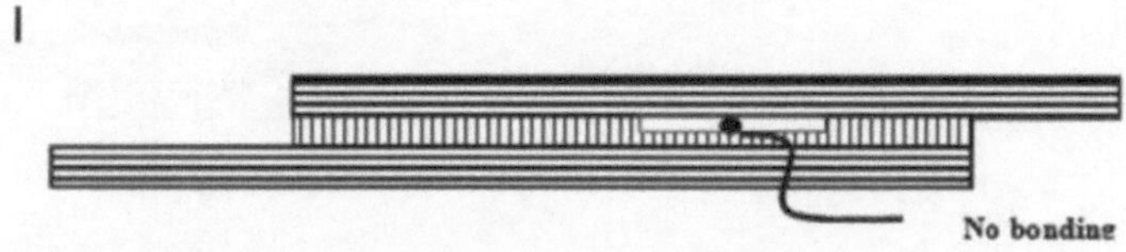

FIGURE 7.2 Disband mode failure initiation.

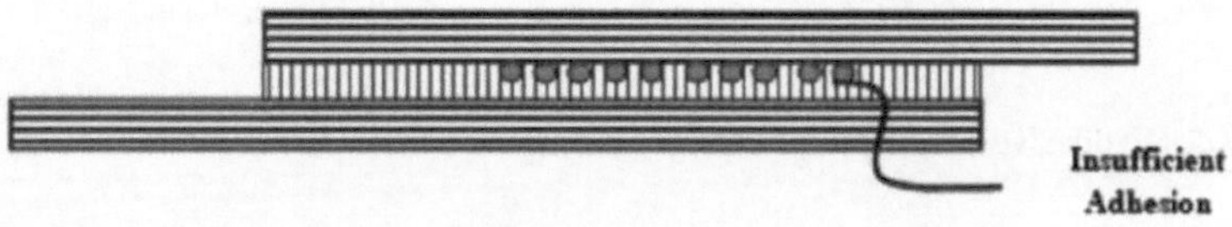

FIGURE 7.3 Weak bond mode failure initiation.

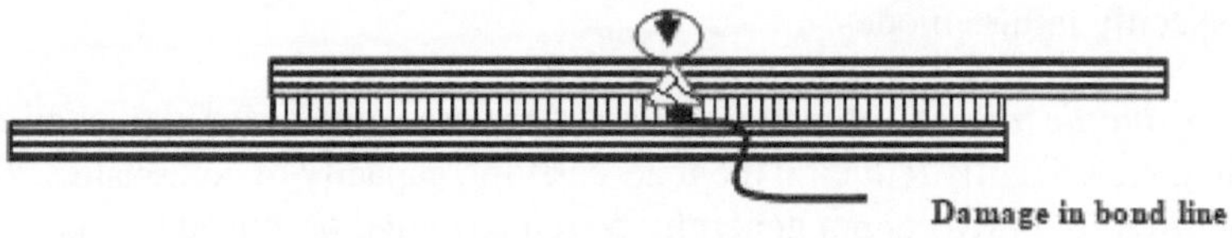

FIGURE 7.4 Impact mode failure initiation.

Impact events in manufacture may cause initial damage of the adherend and the adhesive. NDI can identify damage caused by impact (Figure 7.4).

7.4.3 Joint Configuration Effects

Joints involve discontinuities in geometric structure and material properties as well as large local stress concentrations, and therefore represent one of the most challenging problems in structural design. Joints examined in the literature include strap joints, butt joints, butt strap joints, corner joints, stepped-scarf joints, T-shaped joints, L-shaped joints, double-double joints and tubular lap joints. Common joint configurations studied in the literature are single-lapped, double-lapped, scarf and stepped-lapped joints (Figure 7.5). Adams et al. (1997) investigated a large range of joints. Chamis and Murthy (1991) discussed the step-by-step process for the preliminary design of composite adhesive joints. It includes single, double and step joints with adhesive bonded joints such as lap joints and scarf joints in hot/wet service environments with static and cyclic loads.

7.5 Displaying Methods of Composites Failure

Plasticity, failure criterion and continuum damage mechanics are the three methods used to model the failure behaviours of FRCs.

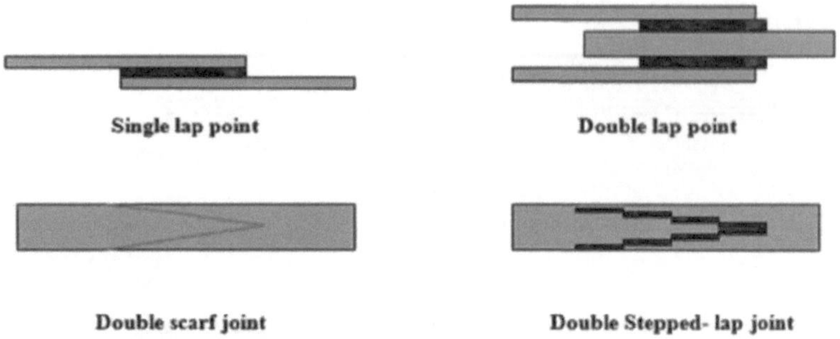

FIGURE 7.5 Adhesive bonded joint configurations.

7.5.1 Failure Criterion Method

Some common failure modes of composites are briefly reviewed in this section. Orifici et al. (2008) provide detailed descriptions of different failure criteria related to these specific failure modes.

(i) *Fibre ductile failure*: Fibre pull-out and fibre breakage occur in this failure situation, significantly reducing the load-carrying capacity of composites. The tensile stress in fibres cannot generally be redistributed or moved to other portions of a structure, resulting in catastrophic failure (Pinho et al. 2005).
(ii) *Fibre compression failure*: This failure mechanism is controlled by the local stability of fibres, which is affected by the local geometry and is less related to the direct strength of composites (Thom 1998).
(iii) *Matrix tensile failure*: The matrix of composites is normally weak under tensile loading. Fibres must be inserted in various orientations to increase load-carrying capability in the matrix tensile direction.
(iv) *Matrix compression failure*: A slant failure surface is commonly found in the form of matrix compression failure.
(v) *Fibre network shear failure*: A matrix will be prone to failure under fibre matrix shear conditions. The failure strength of matrix shearing is determined by compressive stress perpendicular to the direction of the fibres, just as it is in matrix compression failure mode (Puck and Schürmann 1998).

7.5.2 Continuum Damage Mechanics Method

The continuum damage mechanics (CDM) model is the second category observed. Continuous stiffness degradation occurs in fibre-reinforced composite materials. Internal variables are used in CDM models to describe the loss of rigidity. Kachanov (1958) initially created a continuum damage mechanics-based framework to investigate metal creep rupture between local stress in damaged and undamaged configurations.

The internal variables are a key feature of CDM-based techniques. Assigning different damage indicators to different loading conditions helps to distinguish between initial and post-failure behaviours. To include material nonlinearity, these internal variables can alternatively be allocated as a function of strain or strain energy.

7.5.3 Plasticity Method

The plasticity method is usually preferred for ductile composite materials such as graphite/PEEK, aluminium/boron and other thermoplastic composites. Matrix materials like polymers typically exhibit plastic behaviours. The plasticity approach is commonly used in micromechanics composite research (Caporale et al. 2006, Megnis and Varna 2003). This plasticity method also applies at structural level. Barbero and Lonetti (2002) established the plasticity/damage model for a single lamina, which can be built on to characterize the behaviour of polymer matrix composite laminates.

7.5.4 Delamination Modelling

Delamination is an important failure mode exerted inside a composite. The various criteria proposed to predict the initiation and propagation of delamination are utilized in various combinations of transverse tension in a linear or quadratic form of shear and sometimes tension along the direction of the fibres. The maximum stress criterion to predict the beginnings of failure is limited to transverse tensile strength and two shear strengths. Hashin (1980) utilised a quadratic form that includes the transverse tensile stress as well as two shear stresses. Lee (1982) developed a model similar to the maximum stress criterion that included two shear stresses in quadratic form.

7.6 ENVIRONMENTAL PARAMETERS INFLUENCING EFFICIENCY OF BONDED JOINTS

The major environmental impacts that can affect the durability and strength of the joints are moisture absorption and temperature. Figure 7.6 illustrates the primary parameters influencing the durability of adhesively bonded joints, including fire, moisture, UV (ultra violet) radiation and temperature.

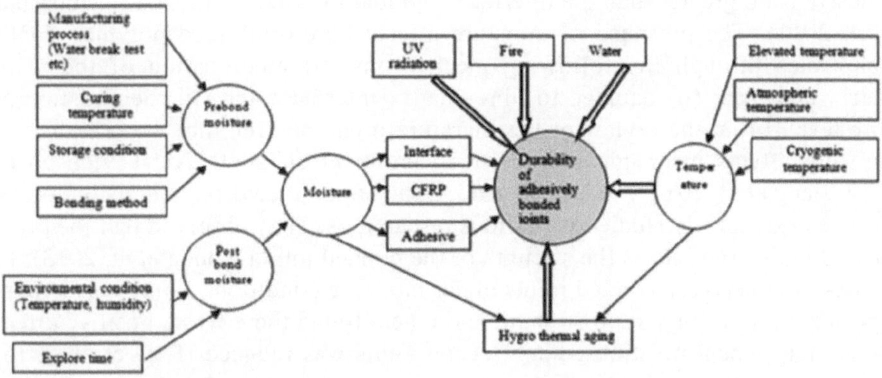

FIGURE 7.6 Natural Parameters Impacting the Strength of Adhesively Fortified Joints.

7.6.1 Moisture in Pre-Bond

Moisture in the pre-bond, which is between the substrates in polymeric composites, is a major issue in adhesive joints and has a direct impact on their performance. The effect of moisture content on the substrate prior to mechanical bonding has been investigated. The substrates absorb moisture in various ways before bonding during the manufacturing process, laminate storing and making of joints. In spite of these structural design aspects, the effect of humidity on the properties of carbon-fibre reinforced polymers (CRPs) and structures is not documented extensively. A few studies of bonded joints for pre-bond moisture content have concluded that joint strength is reduced (Markatos et al. 2013a, Mohan et al. 2013, Budhe et al. 2014, Pantelakis and Tserpes 2013, Tserpes et al. 2014). Budhe and Markatos (2013) concluded that the decrease in fracture toughness could be due to voiding, adhesive plasticization and interfacial adhesion reduction. Markatos et al. (2013a) reported that increasing the substrate drying time, together with a pre-bond and presence of moisture, improves joint toughness, but recovery was not complete. However, the existence of moisture in pre-bond substrates has a significant effect on failure modes, making them subject to multiple cracking and interface failure.

7.6.2 Moisture in Post-Bond

The amount of moisture absorbed by a composite structure during its service life is dependent on the adhesive material used, the bonding method, curing temperature, exposure conditions and time (Carbas et al. 2016, Costa et al. 2016). Generally, moisture affects the adhesive by swelling it, increasing cracks, hydrolyzing it and lowering its glass transition temperature. In addition, moisture content increases ductility, but decreases resin strength and modulus of elasticity. In cyclic conditions, moisture causes loss of ductility and modulus of elasticity (Mubashar et al. 2011). The extraction of water results in plasticization of the resin and the loss of adhesive ductility during cyclic moisture exposure (Ashcroft et al. 2011). The stability of interfacial adhesion is a major factor in long-term durability of adhesive joints in the presence of moisture. The degradation of adhesive is no greater than the interface degradation (Stazi et al. 2015, Mubashar et al. 2011). The presence of moisture affects fibre-reinforced polymer (FRP) composites through the following mechanisms: (i) modification of the resin matrix interface; (ii) damage to fibre–matrix interface; and (iii) degradation at fibre level. From the review of the literature it can be seen that the presence of moisture affects mechanical properties (Sciolti et al. 2010, Aniskevich et al. 2012, Jiang et al. 2014, Zafar et al. 2012). The effect of environmental conditions has been extensively studied by various researchers. They observed that the presence of moisture reduces the strength of the bonded joint (Mohan et al. 2013). In studies of adhesively bonded joints under moisture conditions, some researchers reported a minor impact on strength and others found there was a positive effect on strength when moisture content at the joints was reduced (Reis et al. 2015, Knight et al. 2012).

7.6.3 Temperature

Adhesives are in increasing demand because they have high structural integrity and can withstand high temperature. The stress strain characteristics and toughness of the joint with reference to the adhesive material play a significant role in design. The significant factors with regard to temperature are cured shrink, thermal expansion coefficient and the mechanical characterisation of the adhesive material (Heshmati et al. 2015, Banea et al. 2014). Studies have shown that loss of strength is due to increase and decrease in temperature (Silva et al. 2009, Hu et al. 2013). It is useful for fibre-reinforced polymers to be exposed beyond the glass transition temperature (Wank et al. 2011). Fluctuating temperatures can also cause material de-bonding and weakening at the matrix–bar fibre interfaces (Shenghu et al. 2009). Low temperature results in matrix hardening, embrittlement, bond degradation of the fibre bar matrix and micro cracking (Robert et al. 2010, Wu and Yan 2011). For long-term performance, adhesives should be developed having regard to distinct thermal gradients. However, in respect of bond repair, it should be remembered that components undergo thermal variations in service in structural applications.

7.6.4 Combined Effect of Temperature and Moisture

The combined effect of temperature and moisture is more harmful than individual effects (Liu et al. 2016, Meng et al. 2015, Viana et al. 2016). A survey of literature shows that no extensive work has been done on the humidity effects on the mechanical behaviour of adhesive bonded structures and composite laminates. Sensitivity to moisture absorption will increase under high temperature, causing broken structures. When the adhesive is exposed to high temperature it suffers significant loss of its elastic and tensile properties (Viana et al. 2016).

When aged (hydrothermal) specimens were dried and tested, shear strength due to irreversible changes in materials was observed. While the effects of temperature and moisture on the mechanical characteristics of adhesive and joints are understood, therefore, their combined effect needs to be addressed in view of the ageing issue, and extensive testing of material and joints is required.

7.7 CONCLUSIONS

The following conclusions are drawn from the present study:

- Co-bonding, co-curing and secondary bounding are the most common methods of adhesive bonding.
- Adhesive bonding is found to be one of the most suitable methods for joining composite structures.
- Surface characteristics such as chemical and physical conditions are critical for creating strong interface bonds. Surface preparation is very important in both secondary bonding and composite repairs.
- Potential failure initiation modes such as disband, weak band and impact, together with the effect of joint configurations, are discussed briefly in this

chapter. Composite failure methods include continuum damage mechanics, failure creation, plasticity and delamination modelling.
- Important environmental parameters affecting bonded joint efficiency include temperature, pre-bond moisture and post-bond moisture.

REFERENCES

Abaris, MJH. Adhesive bonding of composites. Lecture notes. Training Inc.

Adams, RD, J Comyn, WC Wake. *Structural Adhesive Joints in Engineering*. Springer, Netherlands (1997).

Aktas A, Polat Z. Improving strength performance of adhesively bonded single lap composite joints. *J Compos Mater* 44 (2010): 2919–2928.

Alessi S, Pitarresi G, Spadaro G. Effect of hydrothermal ageing on the thermal and delamination fracture behaviour of CFRP composites. *Compo Part B Eng* 67 (2014): 145–153.

Aniskevich K, Aniskevich A, Arnautov A, Jansons J. Mechanical properties of pultruded glass fiber-reinforced plastic after moistening. *Compos Struct* 94 (2012): 2914–2919.

Baldan A. Adhesion phenomena in bonded joints. *Int J Adhes Adhe* 38 (2012): 95–116.

Banea MD and da Silva LF. Adhesively bonded joints in composite materials: an overview. *Proc Inst Mec Eng Part L Mate Des App* 223(1) (2009): 1–18.

Barbero E, Lonetti P. An inelastic damage model for fiber reinforced laminates. *J Compos Mater* 36(8) (2002): 941–962.

Bowen RL, Eichmiller FC, Marjenhoff WA, and N. W. Rupp. Adhesive bonding of composites. *J Am Coll Dent*, 56(2) (1989): 10–13.

Budhe S, Rodríguez-Bellido A, Renart J, Mayugo JA, Costa J. Influence of prebond moisture in the adherents on the fracture toughness of bonded joints for composite repairs. *Int J Adhes Adhes* 49 (2014): 80–89.

Campilho RDSG, AMG Pinto, MD Banea, RF Silva and LFM da Silva. Strength Improvement of Adhesively-Bonded Joints Using a Reverse-Bent Geometry. *J Adhes Sci Technol* (2012): 2351–2368.

Caporale A, Luciano R, Sacco E. Micromechanical analysis of interfacial debonding in unidirectional fiber-reinforced composites. *Compos Struct* 84 (31) (2006): 2200–2211.

Carbas RJC, da Silva LFM, Andrés LFS. Effect of carbon black nano particles concentration on the mechanical properties of a structural epoxy adhesive. *Proc IME J Mater Des Appl* (2016).

Chamis, CC, PLN Murthy. Simplified procedures for designing adhesively bonded composite joints. *J Reinf Plast* (1991).

Charalambous G, Allegri G, Hallett SR. Temperature effects on mixed mode I/II delamination under quasi-static and fatigue loading of a carbon/epoxy composite. *Compos Part A Appl Sci* 77 (2015a): 75–86.

Charalambous G, Allegri G, Lander JK, Hallett SR. A cut-ply specimen for the mixed-mode fracture toughness and fatigue characterisation of FRPs. *Compos Part A Appl Sci* 74 (2015b): 77–87.

Coronado P, Argüelles A, Viña J, Mollón V, Vina I. Influence of temperature on a carbon–fibre epoxy composite subjected to static and fatigue loading under mode- I delamination. *Int J Solids Struc* 49 (2012): 2934–2940.

Costa M, Viana G, da Silva LFM, Campilhoand RDSG. Effect of humidity on the mechanical properties of adhesively bonded aluminium joints. *Proc IME J Mater Des* (2016).

da Silva LFM and RDSG Campilho. *Advances in Numerical Modeling of Adhesive Joints*. Springer (2012).

Davidson BD, Kumar M, Soffa MA. Influences of mode ratio and hygro thermal condition on the delamination toughness of a thermoplastic particulate inter layered carbon/epoxy composite. *Comos Part A Appl Sci* 40 (2009): 67–79.

di Ludovico M, Piscitelli F, Prota A, Lavorgna M, Mensitieri G, Manfredi G. Improved mechanical properties of CFRP laminates at elevated temperatures and freeze–thaw cycling. *Construction Build Materials* 31 (2012): 273–283.

Fitton, MD, JG Broughton. Variable modulus adhesives: an approach to optimised joint performance. *Int J Adhes Adhes* (2005).

Gibson RF. A review of recent research on mechanics of multifunctional composite materials and structures. *Compos Struct* 92 (2010): 2793–2810

Hashin Z. Failure criteria for unidirectional fiber composites. *J Appl Mech* 47(2) (1980): 329–334.

Hedges, WL, RA Buyny, MA Boyle, CJ Martin, and KD McVicker. *Development of a self-adhesive prepreg system for sandwich panel applications*. In *International SAMPE Technical Conference* (2004): 3661–3670.

Heshmati M, Haghani R, Al-Emrani M. Environmental durability of adhesively bonded FRP/steel joints in civil engineering applications: state of the art. *Compos Part B Eng* 81 (2015): 259–275.

Higuchi I, Sawa T, Okuno H, Kato S. Three-dimensional finite element analysis of stress response in adhesive butt joints subjected to impact bending moments. *J Adhes* 79 (11) (2003): 1017–1039.

Hollaway LC. A review of the present and future utilisation of FRP composites in the civil infrastructure with reference to their important in-service properties. *Const Build Maters* 24 (2010): 2419–2445.

Hu P, Han X, Li WD, Li L, Shao Q. Research on the static strength performance of adhesive single lap joints subjected to extreme temperature environment for automotive industry. *Int J Adhes* 41 (2013): 119–126.

Im JM, Shin KB, Hwang TK. *Evaluation of mode I, II inter laminar fracture toughness for a filament wound carbon/epoxy composite under low and high temperature. Proceedings of the 16th European conference on composite material.* Spain (2014).

Jahna J, Weebera M, Boehnera J and Steinhilperb R. *Assessment strategies for composite-metal joining technologies – A review. Peer-review under responsibility of the organizing committee of the 26th CIRP Design Conference* (2016): 689–694.

Jiang X, Kolstein H, Bijlaard F, Qiang X. Effects of hygro thermal aging on glass fibre reinforced polymer laminates and adhesive of FRP composite bridge: moisture diffusion characteristics. *Compos Part A Appl Sci* 57 (2014): 49–58.

Johnson WS, Oliver MS. Effect of temperature on Mode I inter laminar fracture of IM7/PETI-5 and IM7/977-2 laminates. *J Compos Mater* 49 (2009): 1213–1219.

Kachanov L. Time of the rupture process under creep conditions. (*J*). *Isv Akad Nauk SSR Otd Tekh Nauk* 8 (1958): 26–31.

Banea MD, da Silva LFM, Campilho RDSG. Effect of temperature on the shear strength of aluminium single lap bonded joints for high temperature applications. *J Adhes Sci Technol* 28 (2014): 1367–1381.

Katnam KB, da Silva LFM, Young TM. Bonded repair of composite aircraft structures: A review of challenges and opportunities. *Progr Aerospace Sci* (2013): 26–42.

Kihara K, Isono H, Yamabe H, Sugibayashi T. A study and evaluation of the shear strength of adhesive layers subjected to impact loads. *Int J Adhes Adhes*. 23 (4) (2003): 253–259.

Kim H, Kayir T, Mousseau SL. Mechanisms of damage formation in transversely impacted glass–epoxy bonded lap joints. *J Compos Mater* 39 (2005): 2039–2052.

Kinloch AJ. *Durability of Structural Adhesives*. 1st ed. Applied Science Publishers Ltd (1983).

Knight GA, Hou TH, Belcher MA, Palmieri FL, Wohl CJ, Connell JW. Hygro thermal aging of composite single lap shear specimens comprised of AF-555M adhesive and T800H/3900-2 adherends. *Int J Adhes* 39 (2012): 1–7.

Kruse T, Korwien T, Heckner S. *Bonding of CFRP primary aerospace structures–crack stopping in composite bonded joints under fatigue on Composite Materials*. In *20th International Conference on Composite Materials, Copenhagen*, 19–24 July 2015.

Lee JD. *Fracture of Composite Materials*. Berlin: Springer (1982) pp. 291–306.

Li, H, Xian, G, Lin, Q, and Zhang, H. Freeze-thaw resistance of unidirectional-fiber-reinforced epoxy composites. *J Appl Polym Sci*, 123(6)(2011): 3781–3788.

Lin J, Hua D, Wang PC, Lu Z, Min J. Effect of thermal exposure on the strength of adhesive-bonded low carbon steel. *Int J Adhes Adhes* 43 (2013): 70–80.

Liu S, Cheng X, Zhang Q, Zhang J, Bao J, Guo X. An investigation of hydro thermal effects on adhesive materials and double lap shear joints of CFRP composite laminates. *Compos Part B Eng*, 91 (2016): 431–440.

Markatos DN, Tserpes KI, Rau E, Brune K, Pantelakis S. Degradation of mode-I fracture toughness of CFRP bonded joints due to release agent and moisture prebond contamination. *J Adhes*, 90 (2013a): 156–173.

Markatos DN, Tserpes KI, Rau E, Markus S, Ehrhart B, Pantelakis S. The effects of manufacturing-induced and in-service related bonding quality reduction on the mode-I fracture toughness of composite bonded joints for aeronautical use. *Compos Part B Eng* 45 (2013b): 556–564.

Megnis M, Varna J. Micromechanics based modeling of nonlinear visco plastic response of unidirectional composite. *Compos Sci Tech* 63(1) (2003): 19–31.

Meng M, Rizvi MJ, Grove SM, Le HR. Effects of hydrothermal stress on the failure of CFRP composites. *Compos Struct* 133 (2015): 1024–1035.

Mohan J, Ivanković A, Murphy N. Effect of prepreg storage humidity on the mixed-mode fracture toughness of a co-cured composite joint. *Compos Part A Appl Sci*, 45 (2013): 23–34.

Mubashar A, Ashcroft I, Critchlow GW, Crocombe AD. A method of predicting the stresses in adhesive joints after cyclic moisture conditioning. *J Adhes*, 87 (2011): 926–950.

Orifici A, Herszberg I, Thomson R. Review of methodologies for composite material modeling incorporating failure. *J Compos Struct* 86(1) (2008): 194–210.

Palmieri FL, Ledesma RI, Dennie JG, Kramer TJ, Lin Y, Hopkins JW Wohl CJ, Connell JW. Optimized surface treatment of aerospace composites using a picosecond laser. *Compos Part B Eng*, 175 (2019).

Pantelakis S, Tserpes KI. Adhesive bonding of composite aircraft structures: challenges and recent developments. *Sci China Phys Mech Astron*, 57 (2013): 2–11.

Park YB, Song MG, Kim JJ, Kweon JH, Choi JH. Strength of carbon/epoxy composite single-lap bonded joints in various environmental conditions. *Compos Struct*, 92 (2010): 2173–2180.

Pinho S T, Dàvila C G, Camanho P P, et al. Failure models and criteria for FRP under in-plane or three-dimensional stress states including shear non-linearity. NASA (2005).

Puck A, Schürmann H. Failure analysis of FRP laminates by means of physically based phenomenological models. *Compos Sci Tech* 58(7) (1998): 1045–1067.

Reis PNB, Soares JRL, Pereira AM, Ferreira JAM. Effect of adherends and environment on static and transverse impact response of adhesive lap joints. *Theor Appl Fract tMech* 80 (2015): 79–86.

Robert M, Roy R, Benmokrane B. Environmental effects on glass fiber reinforced polypropylene thermoplastic composite laminate for structural applications. *Polym Compos*, 31 (2010): 604–611.

Schmid Fuertes TA, Kruse T, Körwien T, Geistbeck M. Bonding of CFRP primary aerospace structures – discussion of the certification boundary conditions and related technology fields addressing the needs for development. *Compos Interfaces* (2015) 22 (8): 795–808.

Sciolti MS, Frigione M, Aiello MA. Wet lay-up manufactured frps for concrete and masonry repair: influence of water on the properties of composites and on their epoxy components. *JCompos Construct* 14 (2010): 823–833.

Shanmugavadivu N. Transverse magnetic mode slab waveguide optical sensor in the presence of conducting interfaces. *Optik-International Journal for Light and Electron Optics*, 178 (2019) 1090–1096.

Shenghu C, Zhis WU, Xin W. Tensile properties of CFRP and hybrid FRP composites at elevated temperatures. *J Compos Mater* 43 (2009): 315–330.

Stazi F, Giampaoli M, Rossi M, Munafò P. Environmental ageing on GFRP pultruded joints: comparison between different adhesives. *Compos Struct* 133 (2015): 404–414.

Teng JG, Yu T, Fernando D. Strengthening of steel structures with fiberreinforced polymer composites. *J Construct Steel Res*, 78: 131–143 (2012).

Thom H. A review of the biaxial strength of fibre-reinforced plastics. *J. Composi Part A Appl Sci Manuf*, 29(8) (1998): 869–886.

Tserpes KI, Maratos DN, Brune K, Hoffmann M, Rau E, Pantelakis S. A detailed experimental study of the effects of pre-bond contamination with a hydraulic fluid, thermal degradation, and poor curing on fracture toughness of composite-bonded joints. *J Adhes Sci Technol*, 28 (2014): 1865–1880.

Vaidya UK, Gautam ARS, Hosur M, Dutta P. Experimental–numerical studies of transverse impact response of adhesively bonded lap joints in composite structures. *Int J Adhes Adhes* 26 (2006): 184–198.

Vassilopoulos, P. *Fatigue and Fracture of Adhesively-bonded Composite Joints Behaviour Simulation and Modelling*. Woodhead Publishing (2015).

Viana G, Costa M, Banea MD, da Silva LFM. "Behaviour of environmentally degraded epoxy adhesives as a function of temperature." *J Adhes* 93 (2017): 95–112.

Viana G, Costa M, Banea MD, da Silva LFM. A review on the temperature and moisture degradation of adhesive joints. *Proc I Mech E Part L J Mater Des Appl* (2016): 1–14.

Wang K, Young B, Smith ST. Mechanical properties of pultruded carbon fibrereinforced polymer (CFRP) plates at elevated temperatures. *Eng Struct*, 33 (2011): 2154–2161.

Weitsman Y. Stresses in adhesive joints due to moisture and temperature. *J Compos Mater* (1977): 11–17

Wu HC, Yan A. Time-dependent deterioration of FRP bridge deck under freeze/thaw conditions. *Compos Part B Eng* 42 (2011): 1226–1232.

Zafar A, Bertocco F, Thomsen JS, Rauhe JC. Investigation of the long term effects of moisture on carbon fibre and epoxy matrix composites. *Compos Sci Technol*, 72 (2012): 656–666.

Zhang Y, Vassilopoulos AP, Keller T. Effects of low and high temperatures on tensile behavior of adhesively-bonded GFRP joints. *Compos Struct* 92 (2010): 1631–1639.

Zhou, Aixi. Joining techniques for fiber reinforced polymer composite bridge deck systems. *Thomas Keller Compos Struct* 69 (2005): 336–345.

8 Vibro-Acoustic Behaviour of a Damaged Honeycomb Core

Mehmet Yetmez, Oguzhan Sen, and Cagrihan Celebi
Zonguldak Bulent Ecevit University, Zonguldak, Turkey

CONTENTS

8.1 SOME STUDIES ON DAMAGED HONEYCOMB CORES

Due to their adjustable material properties, composite materials are increasingly used in many engineering fields (Boudjemai et al., 2012). These structures, which do not compromise their strength despite their weight decrease, find wide usage areas thanks to their advanced load carrying capacity and improved thermal resistance (Zhu et al., 2019). Due to their energy absorption properties sandwich composites are also used in specific transportation applications, as well as in trains and the decks and hulls of ships where vibration and sound problems are intense (Mozafari et al., 2015). Sandwich composite materials have three basic elements: face, core and adhesive (Zenkert, 1995). Thin sandwich panels can replace thick steel structures or thick beams, reducing the corrosion and resistance problems of conventional materials (Shams et al., 2015). In order to bear the core shear load and the faces to bear bending and in-plane loads (Arunkumar et al., 2016), the core can be selected from metals, composite plastics or wood. Because of the nonlinear geometrical and material properties of sandwich materials, studies on honeycomb composites under nonlinear loading are still in progress. Some of these are described below.

Crushed honeycomb core has been modelled using two different finite element methods to understand the behaviour of crushed core structures (Aktay et al., 2008). Honeycombs with different core structures have been damaged at low speed and their

DOI: 10.1201/9781003128861-8

mechanical behaviour examined (Foo et al., 2008). At high-impact energies, the surface is perforated and major damage occurs to the core (Mines et al., 1998). Density plays an important role in the rate of damage to samples (Zhou et al., 2006). In vibro-acoustic testing, two microphones placed perpendicularly to each other have been used to study damaged composite structures (Lin et al., 2011). King et al. (2020) noted a visible decrease in the modulus of elasticity of a sandwich composite with a damaged core structure.

Despite extensive experimental studies, the characteristic core behaviour of a damaged sandwich panel remains an open area for standardization. In order to contribute to this standardization, the cores under impact need to be examined experimentally (Lee, 2004). In the study reported here, the characteristic behaviours of different damaged cores are investigated both experimentally and numerically. The effect of varying the microphone angle is also examined. Results are given in tabular and graphical form.

8.2 A VIBRO-ACOUSTIC METHOD FOR DAMAGED CORES

Four different honeycomb cores are considered, designated specimen 1, specimen 2, specimen 3 and specimen 4 with a damage extent of 6%, 1%, 4% and 5% respectively. Details of the specimens and the impactor nose are given in Table 8.1 and Figures 8.1–8.3. It is obvious that, under impact loading with a round-nosed impactor, the impact energy capacity is considerable for both thick and lightweight core (i.e., specimen 4) materials. The damage history curve for specimen 4 is very important for sound propagation.

Vibro-acoustic tests are conducted in an acoustic chamber with internal dimensions 395 mm × 205 mm × 185 mm. Vibro-acoustic measurements are performed using two microphones (Type46AE, G.R.A.S., Denmark) and a data acquisition system SoMatTM eDAQ-lite with nCode GlyphWorks software (HBM, Germany). Details of the test set-up are given in Figure 8.4. The upper diagram in Figure 8.4 shows both microphones located parallel to the specimen (0°). In the middle diagram the microphone location is at 45° to the specimen, and in the bottom diagram the microphone is set at 90°. The sound source produces constant sound at a frequency of 432 Hz.

TABLE 8.1
Properties of Drop Weight Impact Test Specimens

Specimen	Material	Dimension (mm × mm × mm)	Specimen mass (g)	Distance (m)	Velocity ($m.s^{-1}$)	Impactor mass (g)
1	Aluminium	60 × 87 × 64	70	1	4.42	518
2	Plastic	60 × 87 × 8.5	9	1	4.42	518
3	Aluminium	60 × 87 × 15.7	48	1	4.42	518
4	Plastic	60 × 87 × 51	48	1	4.42	518

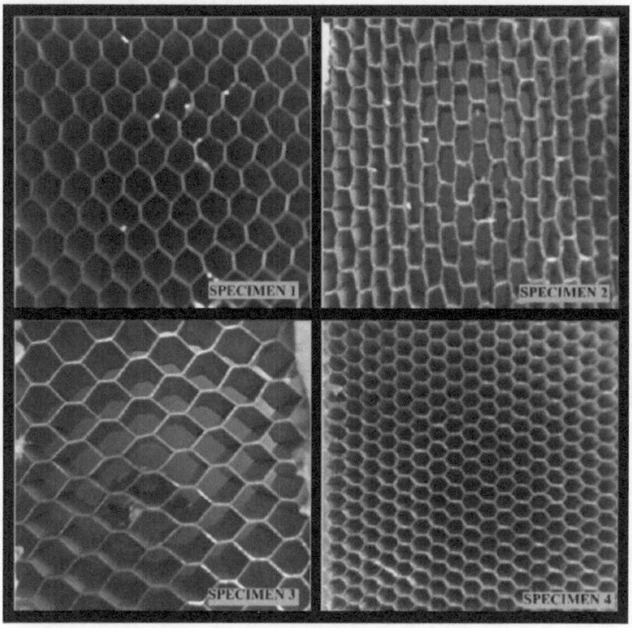

FIGURE 8.1 Geometrical representation of the core specimens.

FIGURE 8.2 Impactor shape for impact testing.

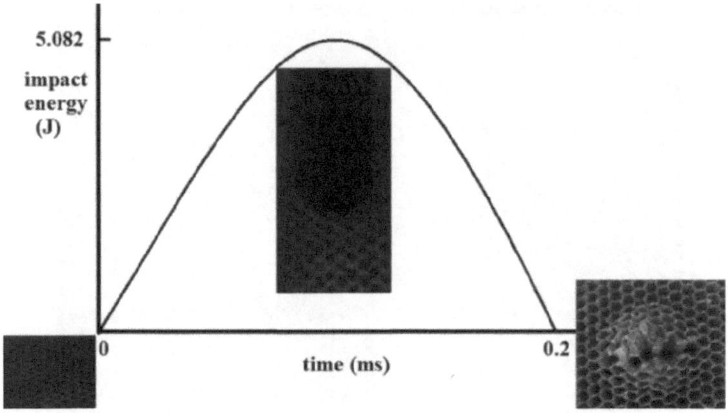

FIGURE 8.3 Damage history of specimen 4.

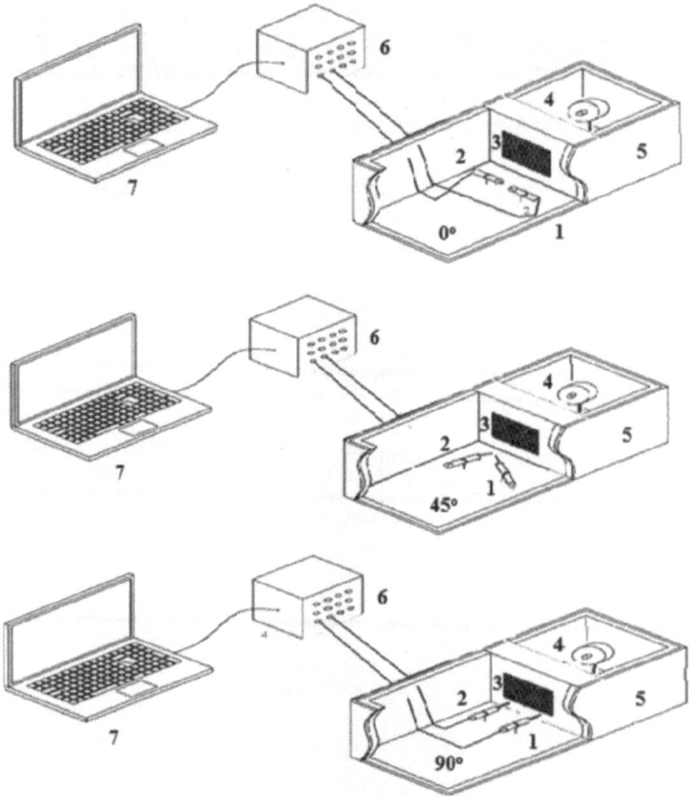

FIGURE 8.4 Vibro-acoustic set-up: (1) microphone at right, (2) microphone at left, (3) specimen, (4) sound source, (5) acoustic chamber, (6) data acquisition system, (7) computer.

8.3 RESULTS FOR CORE'S VIBRO-ACOUSTIC RESPONSES

8.3.1 Experimental Work on Damaged Cores

Figures 8.5–8.8 show the characteristic sound pressure behaviour of undamaged specimens for the three microphone positions. At 0°, increasing the thickness of the aluminum core increases the shifting curve behaviour of sound pressure. Similarly, at 45°, increasing the thickness of the plastic core increases the shifting curve behaviour of sound pressure. One may conclude that the microphone position of 90° is suitable for thin core materials.

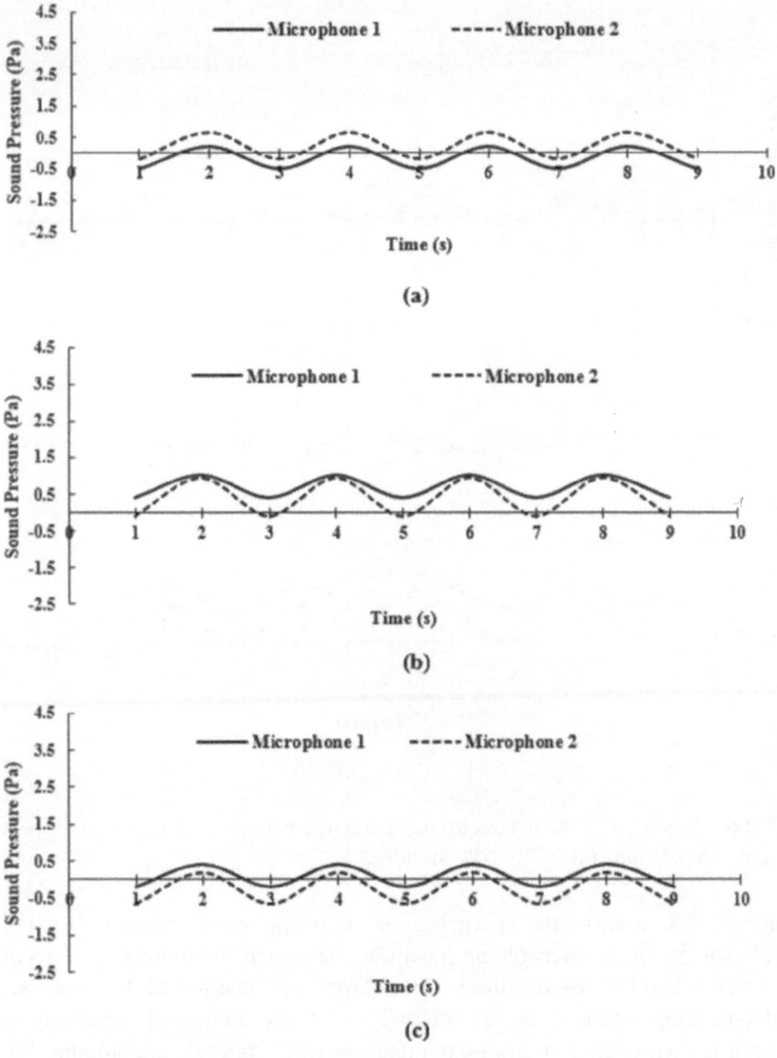

FIGURE 8.5 Variation of characteristic sound pressure behaviour of undamaged specimen 1 at microphone positions (a) 0°, (b) 45°, (c) 90°.

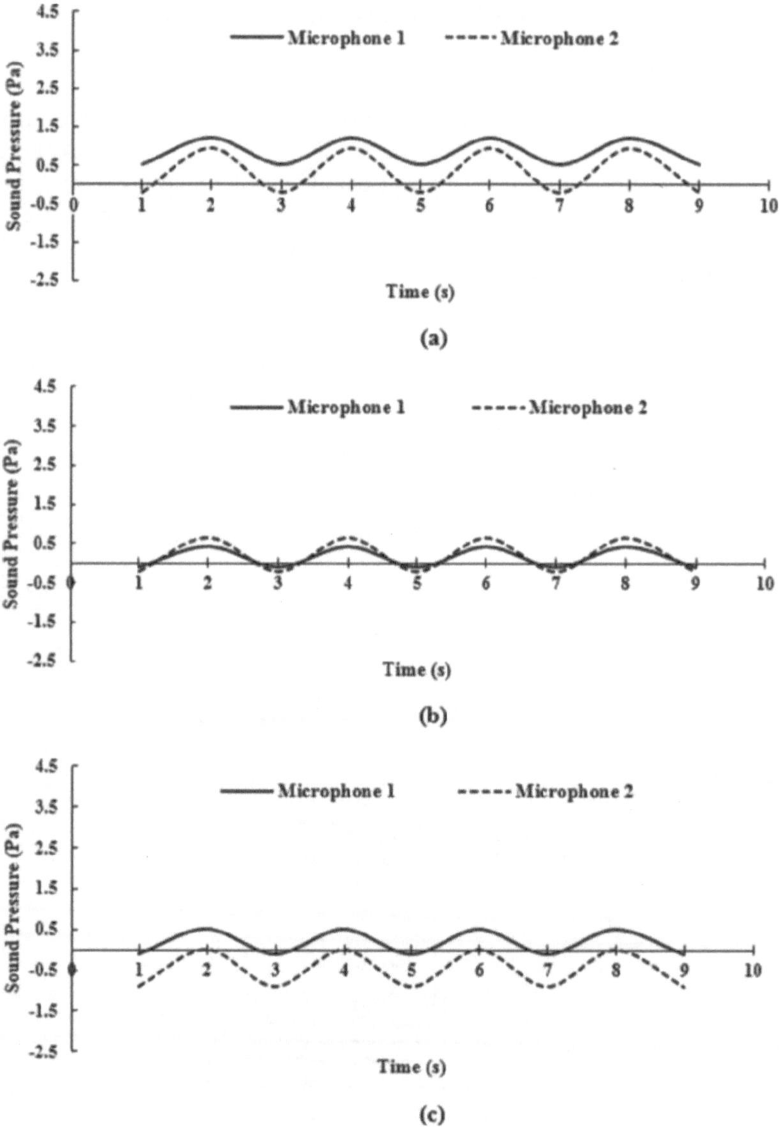

FIGURE 8.6 Variation of characteristic sound pressure behaviour of undamaged specimen 2 for microphone positions (a) 0°, (b) 45°, (c) 90°.

Figures 8.9–8.12 show the characteristic sound pressure behaviour of damaged specimens for the three microphone positions. Although the damage extents are 4% for specimen 3 and 6% for specimen 1, a microphone position of 45° is not suitable for understanding sound pressure behaviour of the damaged aluminum cores. However, it is clearer that microphone positions of 0° and 90° are suitable for thick damaged aluminum cores. There is a close similarity for specimens 2 (damage extent of 1%) and 4 (damage extent of 5%) except for a microphone position of 45°.

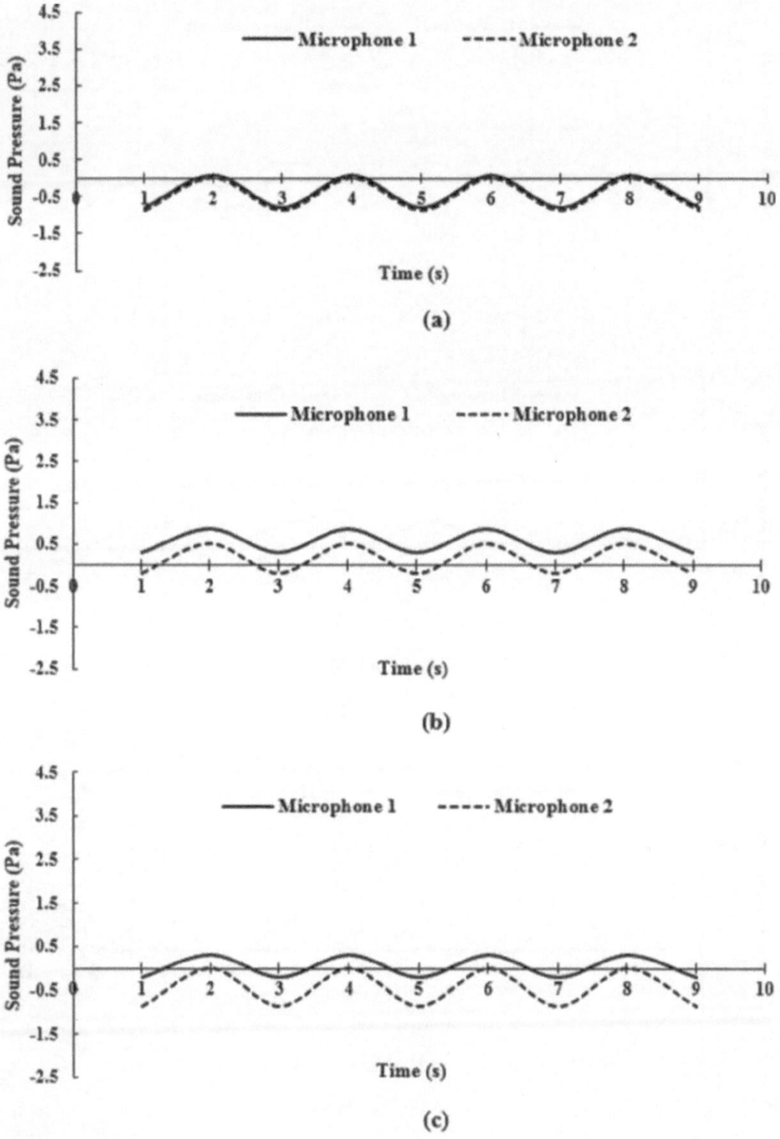

FIGURE 8.7 Variation of characteristic sound pressure behaviour of undamaged specimen 3 for microphone positions (a) 0°, (b) 45°, (c) 90°.

A microphone position of 0° is not suitable for understanding sound pressure behaviour of the damaged plastic cores. On the other hand, it appears that microphone positions of 45° and 90° are suitable for thick damaged plastic cores.

As shown in Figure 8.13, variation of the frequency response function (FRF) of undamaged specimens indicates that (i) increasing the core thickness reveals the

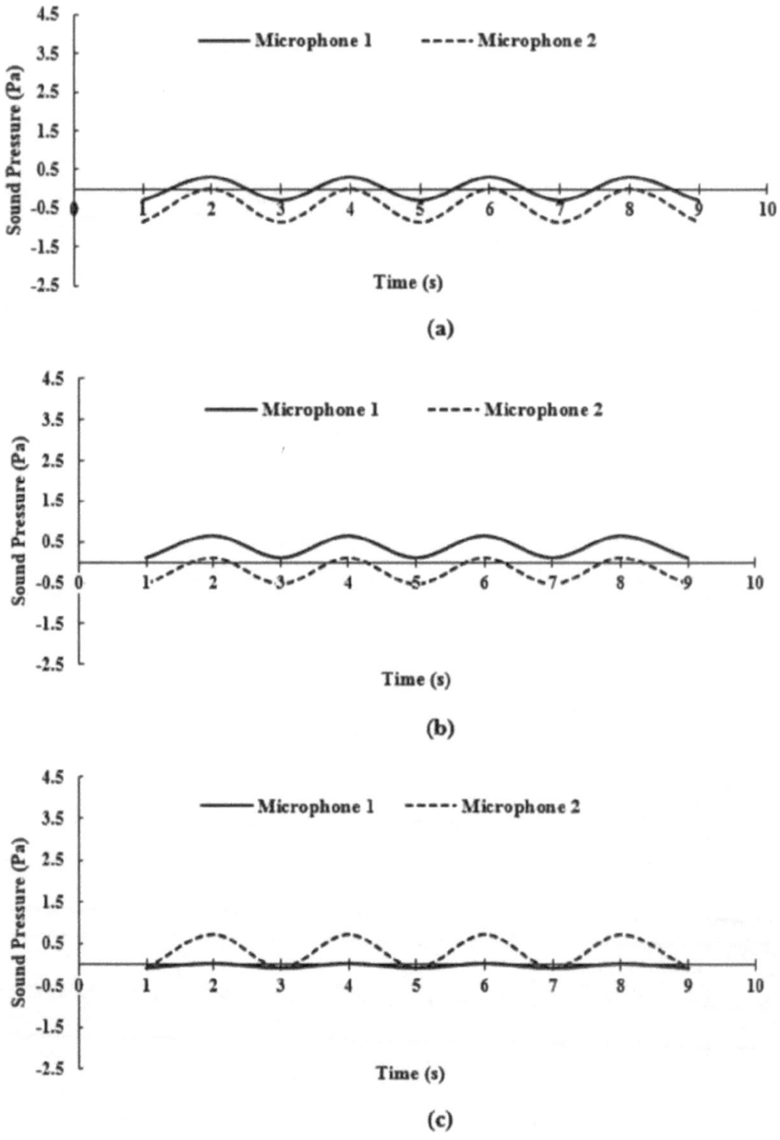

FIGURE 8.8 Variation of characteristic sound pressure behaviour of undamaged specimen 4 for microphone positions (a) 0°, (b) 45°, (c) 90°.

mirror symmetry effect for the aluminium core, (ii) increasing the core thickness increases the microphone position effect for the plastic core.

Figure 8.14 presents the variation of frequency response function (FRF) of damaged specimens, which does not closely resemble that of undamaged specimens. It is obvious that (i) values of FRF are higher than those of undamaged specimens, (ii) microphone effect may not be noted for the thin plastic damaged cores,

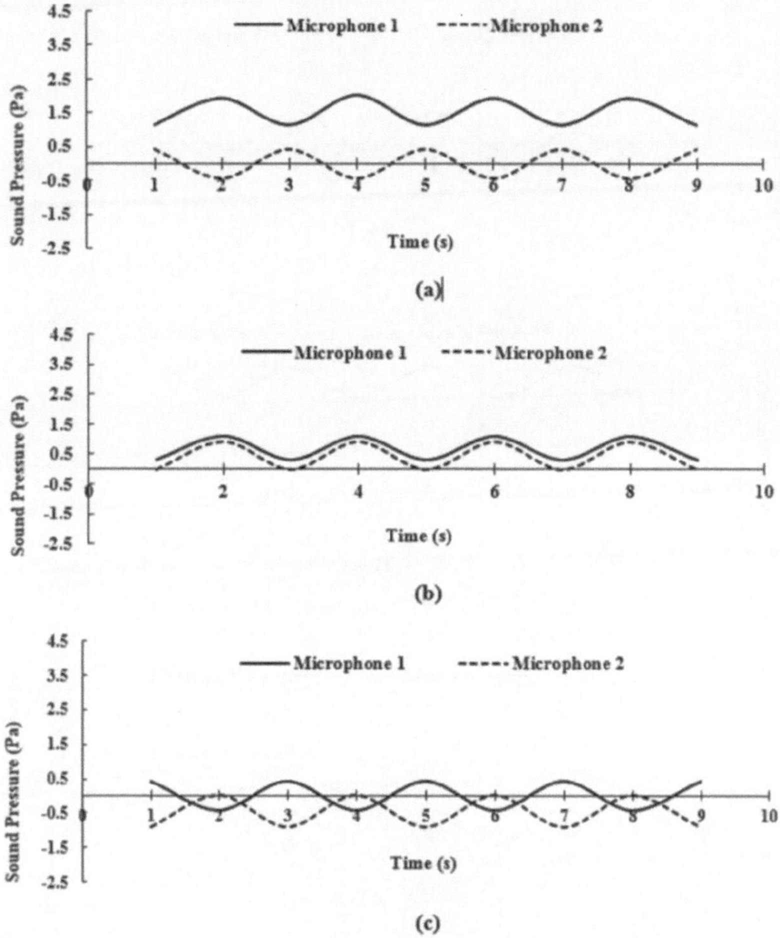

FIGURE 8.9 Variation of characteristic sound pressure behaviour of damaged specimen 1 for microphone positions (a) 0°, (b) 45°, (c) 90°.

(iii) increasing core thickness reveals the mirror symmetry effect for the damaged plastic core, (iv) increasing microphone angle decreases frequency response function (compliance) for the damaged thick aluminum cores.

8.3.2 Numerical Simulation for Damaged Cores

In this study, the general-purpose finite element code ANSYS is used to analyze the vibration characteristics of a damaged honeycomb core. For the finite element models, element SOLID186 is taken into consideration. For each model, a quite significant convergence rate is found using a trial-and-error procedure. Table 8.2 illustrates accuracy in fundamental frequency for specimen 3, examined for various elements. Briefly, the models were as follows: 250028 nodes and 141696 elements for

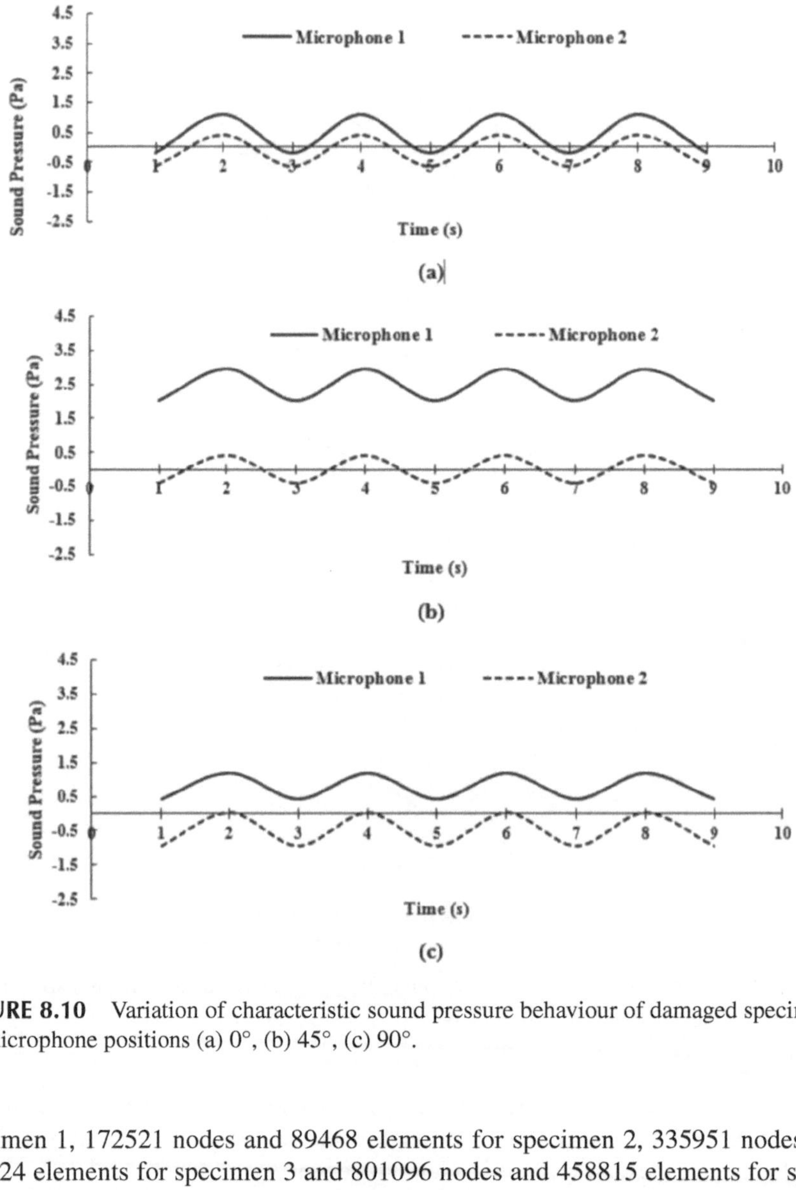

FIGURE 8.10 Variation of characteristic sound pressure behaviour of damaged specimen 2 for microphone positions (a) 0°, (b) 45°, (c) 90°.

specimen 1, 172521 nodes and 89468 elements for specimen 2, 335951 nodes and 174224 elements for specimen 3 and 801096 nodes and 458815 elements for specimen 4. Figure 8.15 gives the accurate representation of complex core geometry.

In this study, both dynamic and linear finite element processes are taken into account. Numerical results show that the highest difference occurs in the damaged thin plastic core.

Fundamental frequency difference between the experimental and numerical results in Figure 8.16 indicates that (i) a finite element model with dynamic and linear finite element processes is moderately suitable for plastic damaged cores, (ii) a dynamic, nonlinear finite element solving process with realistic interface boundary conditions should be considered especially for metallic cores.

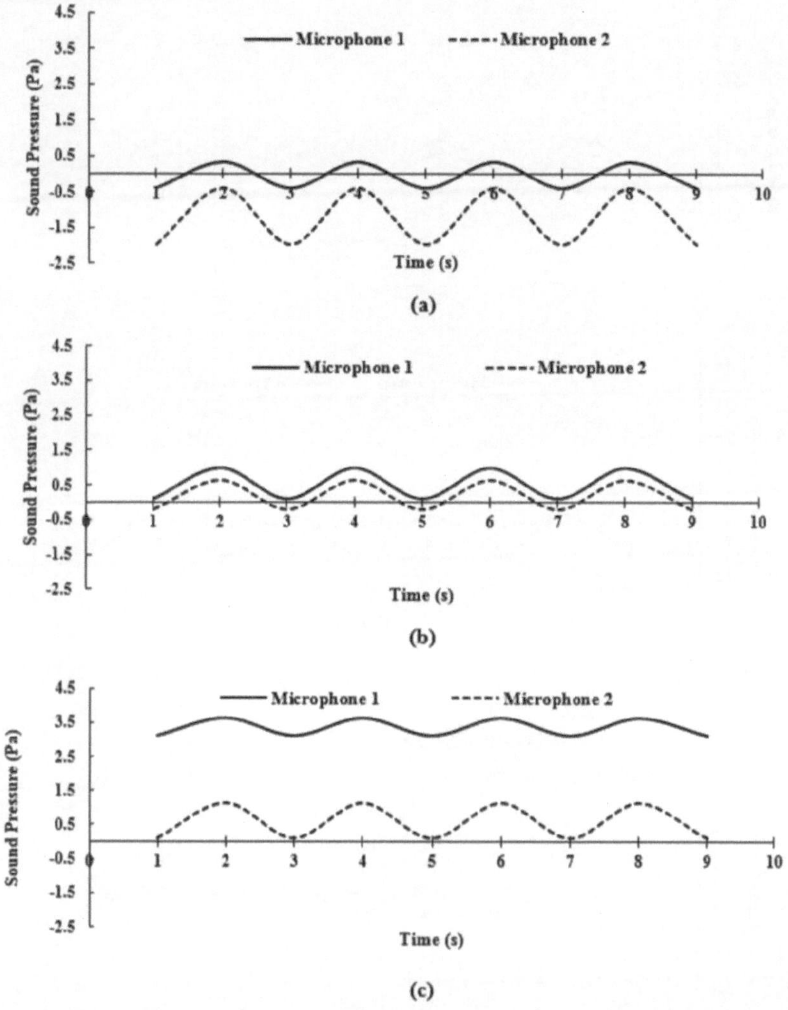

FIGURE 8.11 Variation of characteristic sound pressure behaviour of damaged specimen 3 for microphone positions (a) 0°, (b) 45°, (c) 90°.

8.4 A BRIEF CONCLUSION FOR THE DAMAGED CORES

Vibro-acoustic literature mainly focuses on dimensions and design of stability or damage-related parameters of composite structures such as honeycomb composites and honeycomb core. Understanding the damage region and its topological properties are the main objectives (Patil and Reddy, 2020). To assess energy absorption capacity and structural sustainability under dynamic loading, characteristic damaged behaviours of different cores were investigated to establish whether microphone position angle is a key point for the correct evaluation of dynamic characteristics. In further studies, more suitable numerical models are to be investigated or improved, addressing both geometrical and material nonlinearity conditions.

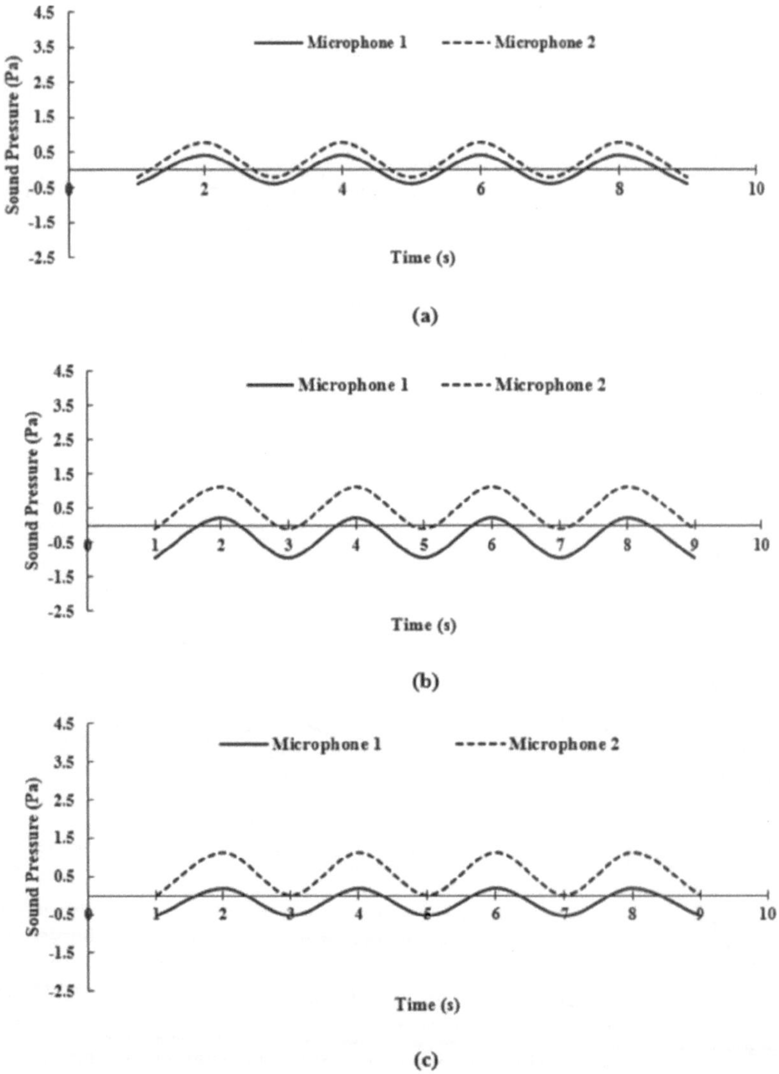

FIGURE 8.12 Variation of characteristic sound pressure behaviour of damaged specimen 4 for microphone positions (a) 0°, (b) 45°, (c) 90°.

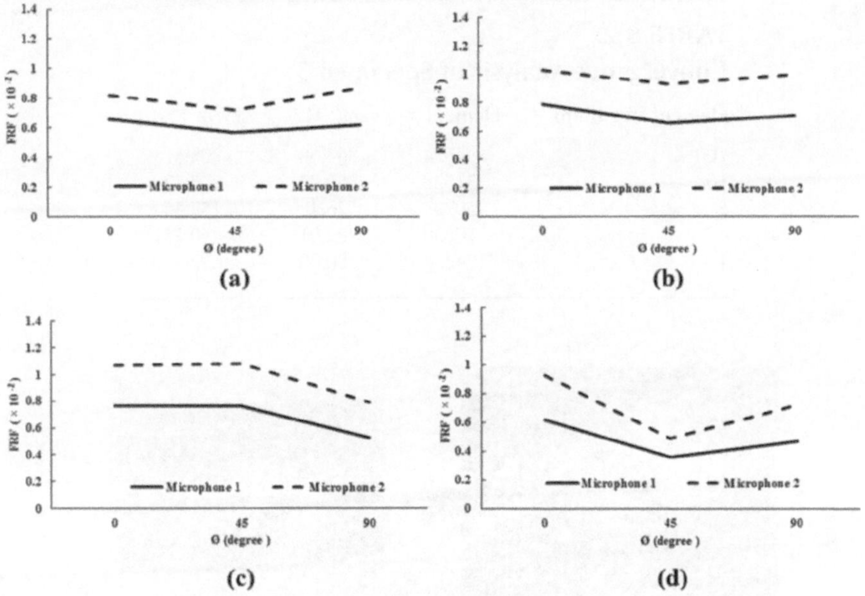

FIGURE 8.13 Variation of frequency response function (FRF) of undamaged specimens: (a) specimen 1, (b) specimen 2, (c) specimen 3 and (d) specimen 4.

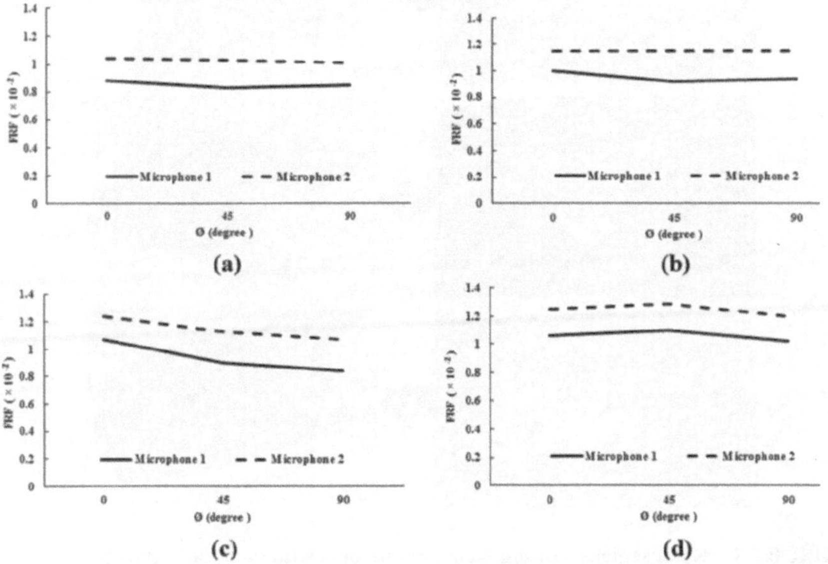

FIGURE 8.14 Variation of frequency response function (FRF) of damaged specimens: (a) specimen 1, (b) specimen 2, (c) specimen 3 and (d) specimen 4.

TABLE 8.2
Convergence Analysis of Specimen 3

Element size (mm)	Element #	ω_1 (Hz)	Error (%)
100	4654	864.68	208.81
10	31628	860.92	207.47
5	50723	726.48	159.46
4	70830	282.01	0.72
1	174224	280.00	—

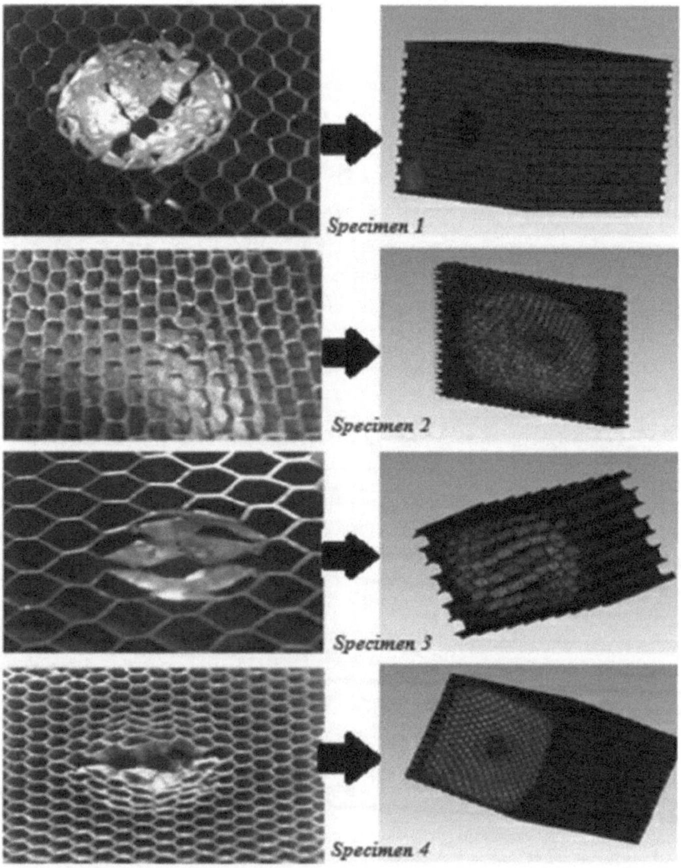

FIGURE 8.15 Representation of finite element models with their damaged cores.

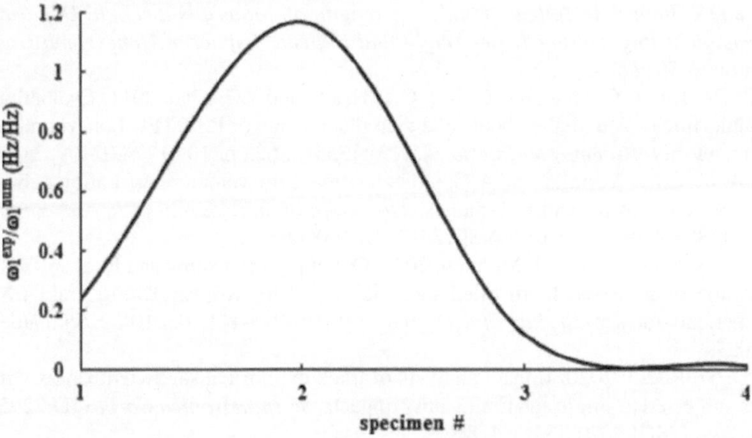

FIGURE 8.16 Fundamental frequency difference between experimental and numerical results.

8.5 CONCLUSION

Vibro-acoustic literature mainly focuses on dimensions and design of stability or damage related parameters of composite structures such as honeycomb composites and honeycomb core. Damage region and understanding its topological properties is the main object on this issue (Patil and Reddy 2020). For the sake of energy absorption capacity and structural sustainability under dynamic loading, characteristic damaged behaviours of different cores are to be investigated to examine whether microphone position angle is the key point to evaluate dynamic characteristics correctly. For the further studies, more suitable numerical models are to be investigated or improved according to both geometrical and material nonlinearity conditions.

REFERENCES

Aktay, L., A.F. Johnson, and B.H. Kröplin. 2008. Numerical modelling of honeycomb core crush behaviour. *Engineering Fracture Mechanics* 75(9): 2616–2630. doi:10.1016/j.engfracmech.2007.03.008.

Arunkumar, M.P., M. Jagadeesh, J. Pitchaimani, K.V. Gangadharan, and M.C. Lenin Babu. 2016. Sound radiation and transmission loss characteristics of a honeycomb sandwich panel with composite facings: effect of inherent material damping. *Journal of Sound and Vibration* 383: 221–232. doi:10.1016/j.jsv.2016.07.028.

Boudjemai, A., R. Amri, A. Mankour, H. Salem, M.H. Bouanane, and D. Boutchicha. 2012. Modal analysis and testing of hexagonal honeycomb plates used for satellite structural design. *Materials and Design* 35 (March): 266–725. doi:10.1016/j.matdes.2011.09.012.

Foo, C.C., L.K. Seah, and G.B. Chai. 2008. Low-velocity impact failure of aluminium honeycomb sandwich panels. *Composite Structures* 85(1): 20–28. doi:10.1016/j.compstruct.2007.10.016.

King, W.T., W.E. Guin, J. Brian Jordon, M.E. Barkey, and P.G. Allison. 2020. Effects of honeycomb core damage on the performance of composite sandwich structures. *Journal of Composite Materials* 54(16): 2159–2171. doi:10.1177/0021998319890656.

Lee, H. 2004. *Imperial College Londan Strength of Materials Section Department of Mechanical Engineering Drop-Weight and Ballistic Impact of Honeycomb Composite Sandwich Structures.*

Lin, J.H., C.M. Lin, C.C. Huang, C.C. Lin, C.T. Hsieh, and Y.C. Liao. 2011. Evaluation of the manufacture of sound absorbent sandwich plank made of PET/TPU honeycomb grid/PU foam.*JournalofCompositeMaterials*45(13):1355–1362.doi:10.1177/0021998310381438.

Mines, R.A.W., C.M. Worralls, and A.G. Gibson. 1998. Low velocity perforation behaviour of polymer composite sandwich panels. *The International Journal of Impact Engineering* 21(10):855–879. doi: 10.1016/S0734-743X(98)00037-2.

Mozafari, H., S. Khatami, and H. Molatefi. 2015. Out of plane crushing and local stiffness determination of proposed foam filled sandwich panel for Korean Tilting Train EXpress - Numerical study. *Materials and Design* 66(PB): 400–411. doi:10.1016/j.matdes.2014.07.037.

Patil, S., D. M. Reddy 2020. Impact analysis of thick cylindrical sandwich panels with foam core subjected to single and multi-mass impacts. *Scientia Iranica Corpus* ID:229567644. doi: 24200/SCI.2020.55341.4180.

Shams, A., A. Stark, F. Hoogen, J. Hegger, and H. Schneider. 2015. Innovative sandwich structures made of high performance concrete and foamed polyurethane. *Composite Structures* 121 (March): 271–279. doi:10.1016/j.compstruct.2014.11.026.

Shreekant P., D. Mallikarjuna Reddy. 2020. Impact damage assessment in carbon fiber reinforced composite using vibration-based new damage index and ultrasonic C-scanning method. *Structures* 28: 638–650. doi:10.1016/j.istruc.2020.09.011

Zenkert, D. 1995. *An Introduction to Sandwich Construction.* Engineering Materials Advisory Services.

Zhou, G., M. Hill, J. Loughlan, and N. Hookham. 2006.Damage characteristics of composite honeycomb sandwich panels in bending under quasi-static loading. *Journal of Sandwich Structures and Materials* 8(1): 55–90. doi:10.1177/1099636206056888.

Zhu, L., W. Liu, H. Fang, J. Chen, Y. Zhuang, and J. Han. 2019. Design and simulation of innovative foam-filled lattice composite bumper system for bridge protection in ship collisions. *Composites Part B: Engineering* 157 (January): 24–35. doi:10.1016/j.compositesb.2018.08.067.

9 Synthesis of Green Hybrid Composite Films for Packaging Applications

Comparative Study with Conventional Materials

Deepak Kohli
National Institute of Technology, Jalandhar, Punjab, India

CONTENTS

DOI: 10.1201/9781003128861-9

9.1 INTRODUCTION

The plastic being used in today's plastic industry is of low cost and good aesthetic qualities. Nearly 42% of plastic is used by the packaging sector (Fortunati et al., 2013). Packaging items made from non-biodegradable polymers are discarded after consumption, causing environmental pollution and damage to wildlife. Plastic waste of 12,000 metric tons will be generated annually in the environment by 2050 (Geyer et al., 2017). There has been great demand to replace non-biodegradable polymers with biodegradable polymers and derivatives that can compete on degradability, cost and physical properties (Sanjay et al., 2018a & Sanjay et al., 2018b).

PVA is a hydrophilic polymer which forms hydrogen bonds with natural fibres. PVA is widely used in the production of packaging and mulch films. However, PVA has lower biodegradability then other biodegradable polymers due to carbon–carbon linkages in the backbone. It is therefore desirable to add reinforcing fillers to the PVA matrix.

Natural fibres from agro-waste such as wood, jute, wheat straw, rice husk and barley husk have been utilized as a reinforcing agent in polymer composites. Natural fibres consist of cellulose, hemicelluloses and lignin. The main component of natural fibre is cellulose which gives strength to the composite. Polymer nanocomposites reinforced with nanoclay have gained significant attention due to their superior physicochemical and barrier properties. The most abundant clay mineral on earth is kaolinite, which is found in sedimentary rocks and soils. Barley husk, a lignocellulosic agro-waste material containing 66% cellulose, can replace synthetic fillers (Ighwela et al., 2012).

While single reinforced composites provide improvement in respect of a few specific properties, they lack other properties. To achieve the desired properties, therefore, the development of hybrid composites is necessary. Recent studies have examined the hybridization of natural fibre with another natural fibre, synthetic fibre and nano fillers using non-biodegradable polymer matrix. Hybrid composites were synthesized using hemp and glass fibre in a polypropylene matrix. It was found that the incorporation of hemp and glass fibre enhanced the tensile properties of the composites. Similarly, tensile strength and modulus of polylactic acid filled with sisal and banana fibre improved by 21% and 40% (Asaithambi et al., 2014). Another study by Hossen et al. (2015) was done on polyethylene hybrid composites reinforced with jute fibre and montmorillonite clay. It was reported that the mechanical properties increased significantly with the addition of 2% clay compared to the composites without nanoclay.

Hybrid composites find applications in packaging, biomedical devices and automotive industries among others (Ali and Ahmad, 2012, Salem et al., 2017, Nayak et al., 2009, Tan et al., 2014, Tan et al., 2015 & Haq et al., 2008). Hybrid films from the present study were used for waste collection bags, hospital waste bags and carrier bags, etc. To date there have been very few studies on the application of kaolinite clay in hybrid films. Hybrid composite films using a biodegradable PVA matrix reinforced with agro-waste (barley husk) and nanoclay (kaolinite) could enhance the properties of films and provide a new crop of stronger, high-barrier and environmentally compatible packaging materials (Melo et al., 2010).

The aim of the present study is to develop composite and hybrid films by incorporating natural fillers and nano clay in the biodegradable PVA matrix. In the study, composite films were synthesized using natural husk barley (NBH), kaolinite clay (KC) and poly vinyl alcohol (PVA). The effect of clay and barley husk on the various properties of the composites was examined.

9.2 MATERIAL AND METHODS

9.2.1 Materials

PVA was supplied by Loba Chemie Pvt. Ltd. (Mumbai, India). Kaolinite clay was supplied by Sigma-Aldrich and sulphuric acid was obtained from Loba Chemicals. Barley was obtained from a nearby market.

9.2.2 Pretreatment of NBH

Barley husk was removed using a grinder. After separation the husk was washed with distilled water. Finally, the barley husk was ground in a grinder and passed through a 100-mesh sieve (size: 0.152 mm). The ground barley husk was stored in an airtight container.

9.2.3 Preparation of Films

PVA (10 gm) was added in distllled water and stirred by stirrer for more than one hour at 90°C to obtain a 10% (weight/volume) solution. Thereafter, various amount (5–25%) of NBH and KC were added in PVA solution followed by continous stirring for half an hour at 80 °C. Then similar weights of the solutions were cast on Petri dishes. The formulations of poly vinyl alcohol/barley husk/kaolinite clay-based composite films are shown in Table 9.1.

9.3 MEASUREMENTS

9.3.1 Scanning Electron Microscopy

Scaning electron microscopy of PVA film, composite and hybrid films was recorded using a scanning electron microscope (JEOL JSM-6100, Tokyo, Japan) with a magnification range of 200 X to 4000 X.

9.3.2 Water Absorption

Film specimens (25 mm × 25 mm) were dried at 50 °C for 24 h in an oven. After drying, the samples were immersed in distilled water at 25 °C for 24 h. Water uptake was calculated as follows:

$$\text{Water uptake}(\%) = \frac{W_1 - W_2}{W_2} \times 100 \tag{9.1}$$

W_1 is the wet film weight and W_2 represents the dry film weight.

TABLE 9.1
Formulations of the Composite Films

S.NO	Composite film code	PVA (wt. %)	Kaolinite clay (KC wt. %)	Natural barley husk (NBH wt. %)
1	PVA/0 KC/0 NBH	100	0	0
2	PVA/5 KC/0 NBH	95	5	0
3	PVA/15 KC/0 NBH	85	15	0
4	PVA/25 KC/0 NBH	75	25	0
5	PVA/0 KC/5 NBH	95	0	5
6	PVA/0 KC/15 NBH	85	0	15
7	PVA/0 KC/25 NBH	75	0	25
8	PVA/5 KC/5 NBH	90	5	5
9	PVA/5 KC/15 NBH	80	5	15
10	PVA/5 KC/25 NBH	70	5	25
11	PVA/15 KC/5 NBH	80	15	5
12	PVA/15 KC/15 NBH	70	15	15
13	PVA/15 KC/25 NBH	60	15	25
14	PVA/25 KC/5 NBH	70	25	5
15	PVA/25 KC/15 NBH	60	25	15
16	PVA/25 KC/25 NBH	50	25	25

9.3.3 Mechanical Properties

Tensile tests were performed with a UTM machine (EZ 20 kN Lloyd Instruments, UK) in accordance with ASTM D638 (Type IV). The samples were cut in a dumbbell shape with a cross-head speed of 20 mm/min.

9.3.4 Light Transmittance

Light transmittance of the films was evaluated through a UV–vis spectrophotometer (PerkinElmer Inc.) with wavelengths of 300 nm to 900 nm. The films were cut in rectangular pieces and directly fixed between two spectrophotometer magnetic cell holders.

9.3.5 Natural Soil Burial

For biodegradation studies, film samples (4 cm × 10 cm) were incorporated in a series of buckets (15 cm deep) containing soil. The samples were regularly moistened with sewage water collected from a sewage treatment plant. The samples were removed after 15 days, cleaned with water and dried.

$$\text{Weight loss}(\%) = \frac{W_1 - W_2}{W_1} \times 100 \tag{9.2}$$

where W_1: weight of existing film and W_2: weight of sample at any time t.

9.4 RESULTS

9.4.1 Mechanical Properties

Tensile strength of PVA increased by 56% and 39.79% with the addition of 5 and 25 wt % KC, respectively (Figure 9.1a). However with the addition of KC, a negative trend was obsèrved for tensile strength. Results indicated that at lower loading, the particles of KC tend to attach effectively onto the PVA matrix, enhancing the tensile properties (Mbey et al., 2012). However, an excessive amount of KC resulted in agglomeration and poor distribution of particles. The elongation of neat PVA film was 271.57%. Elongation increased to 452.25% with the incorporation of KC (5%). This behaviour indicated that incorporating a small amount of KC had a lubricating effect.

The tensile strength of poly vinyl alcohol films was 19.4 MPa. For PVA/5NBH film, tensile strength increased by 31.7% compared to PVA film, and for PVA/10 NBH film it increased by 44.79%. However, with the addition of 10–25% NBH the tensile strength of PVA film reduced by 58.3%. Elongation was 271.57% for PVA film, and for PVA/5NBH composite film, it was reduced to 243.09%. The elongation percentage further decreased with an increase in NBH content.

Figure 9.1b shows the influence of NBH and KC on mechanical properties of PVA/KC/NBH hybrid films. For PVA film, tensile strength increased by 31.7% with the incorporation of 5% NBH in PVA/5NBH composite film. With 5% loading of KC in hybrid composite film (PVA/5NBH/5KC), the tensile strength was enhanced by 14.79% as compared to PVA/5NBH film. For PVA/5NBH/15KC film containing 15% KC, tensile strength increased to 31.22MPa. However for higher KC content, the reverse trend was observed. The PVA film showed an elongation of 271% which

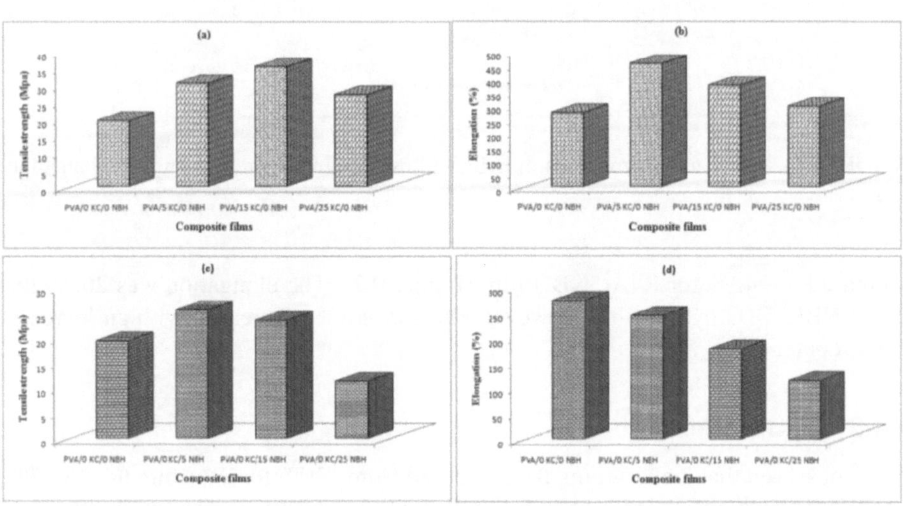

FIGURE 9.1 Mechanical properties: (a) tensile strength of PVA/KC films, (b) elongation (%) of PVA/KC films, (c) tensile strength of PVA/NBH films and (d) elongation (%) of PVA/NBH films.

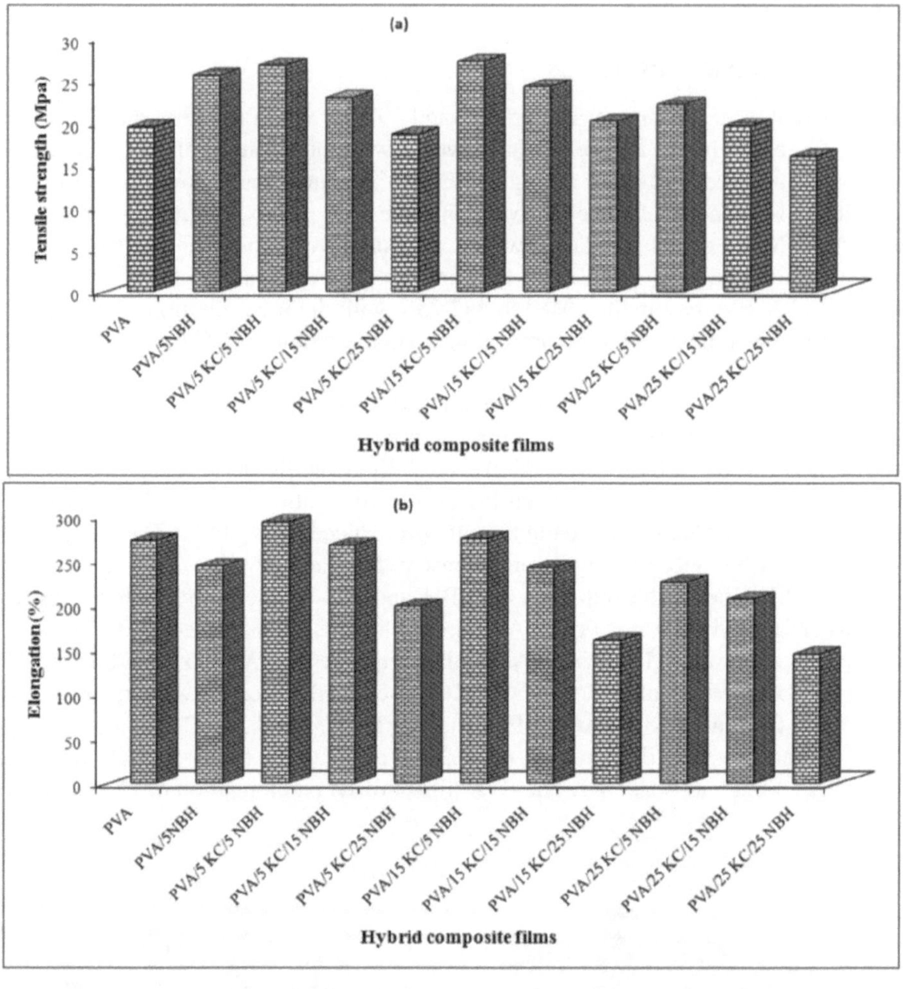

FIGURE 9.2 Mechanical properties of hybrid films: (a) tensile strength and (b) elongation (%).

decreased to 243% for PVA/5NBH film (Figure 9.2). The elongation was 262% for PVA/5NBH/5KC hybrid film. However, elongation (%) decreased with an increase in KC content.

9.4.2 Water Uptake

It was observed that the water uptake decreased from 130% for PVA film to 125.02% for PVA/15KC/0NBH. This behaviour indicated that the KC increased the tortuous diffusion path and also increased the PVA–clay bondings (Hossain et al., 2012).

Figure 9.3b shows the water uptake (%) of NBH-reinforced composite films. After 6 days, the water uptake of PVA film was reported to be 116%. The composite

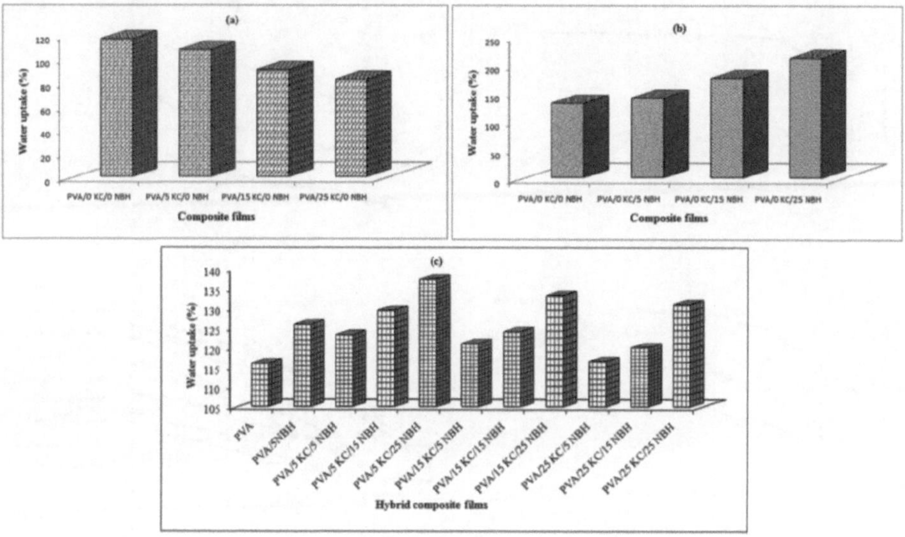

FIGURE 9.3 Water uptake studies: (a) PVA/KC composite films, (b) PVA/NBH composite films and (c) PVA/KC/NBH hybrid composite films.

film (PVA/5NBH) containing 5% NBH showed water uptake of 126% after 6 days. Results indicated that increasing the NBH loading from 5% to 25% increased the water uptake to 184.5% (after 6 days).

Water uptake of PVA and PVA/5NBH composite film was 52% and 64% after 1 day, increasing to 116% and 126% after 6 days. For PVA/5KC/5NBH hybrid film, water uptake reduced by 5.19% compared to PVA/5NBH film after 6 days, whereas for PVA/5NBH/25KC film, water uptake was 110.5% after 6 days and was lower than PVA film. When the barley husk content was further increased up to 25% in the hybrid film at 5% KC loading, the water uptake increased significantly. This was due to the increase in OH groups which increased the hydrophilicity of the films. With KC loadings, the water uptake reduced due to enhanced PVA–clay interactions via formation of hydrogen bonds. The presence of impermeable KC layers within the PVA matrix increased the tortuosity of the diffusion path for the water molecules and also reduced the length of the free water diffusion route (Fortunati et al., 2012).

9.4.3 Optical Properties

The optical transparency (T%) of the films at different wavelengths is shown in Figure 9.4. PVA has 92% transmittance at 700 nm. When 5 wt % KC is added to the PVA matrix, the transmittance is reduced to 75%. PVA/5 KC/0NBH films showed higher transmittance than PVA/15 KC/0NBH film due to better distribution of clay particles at low content.

It was observed that the transmittance reduced when NBH was incorporated into the PVA matrix (Figure 9.4c). For PVA/5 NBH, the transmittance reduced to 71%.

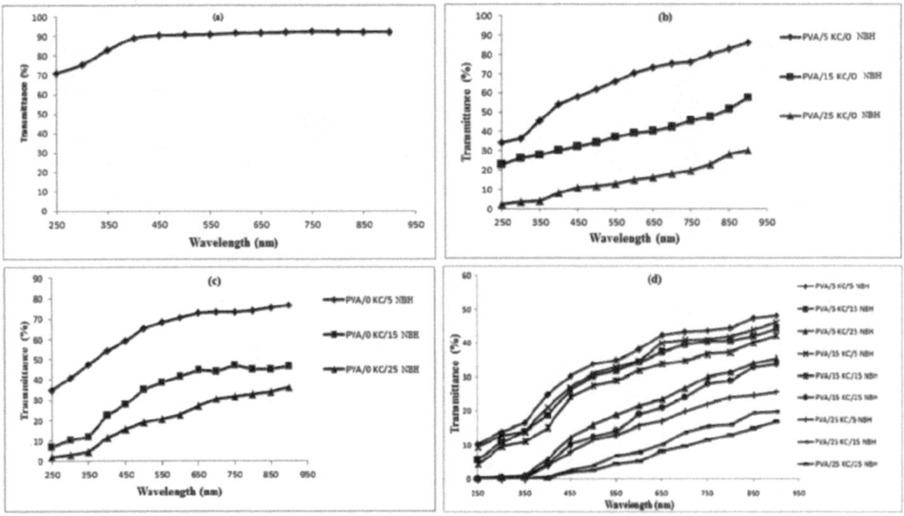

FIGURE 9.4 Transmittance (%) of the composite films: (a) PVA film, (b) PVA/KC films, (c) PVA/NBH films and (d) PVA/KC/NBH hybrid composite films.

In the composite film containing 25% NBH, the transmittance reduced to 26.7% at 700 nm. This reduction in transmission was mainly due to the absorption and restriction of light by barley husk (Cheng et al., 2010; Tagreed, 2013).

The light transmittance (%) for PVA/NBHs/KC hybrid composite films at different wavelengths is presented in Figure 9.4d. It was observed that the transmittance (%) of films decreased with increase in filler loading, but increased with increase in wavelength. The PVA film showed the highest transmittance of 92% at 700 nm. PVA/5NBH composite film showed transmittance of 71% which was less than PVA film. Transmittance reduced to 51% for PVA/5NBH/5KC film. The transmittance reduced further with the incorporation of clay from 5% to 25% in the hybrid. Hybrid film (PVA/25NBH/25KC) containing 25% NBH and 25% KC had the lowest transmittance of 10.5% at 700 nm.

9.4.4 Soil Burial Test

Composite films were exposed to soil for 180 days under environmental conditions. The composite films decreased in size and appeared hard and brittle after soil burial. The weight loss of the KC-filled PVA films was lower than the unfilled PVA matrix and rapidly decreased with increasing KC content, which was due to presence of kaolinite clay particles decreasing the penetration of microorganisms into the film (Hossain et al., 2012).

Figure 9.5b shows the variation in weight loss of PVA/NBH composite films with time. The soil burial test was conducted for 180 days. The unfilled PVA attained a maximum weight loss of 20% within 30 days. However, all the composite films had

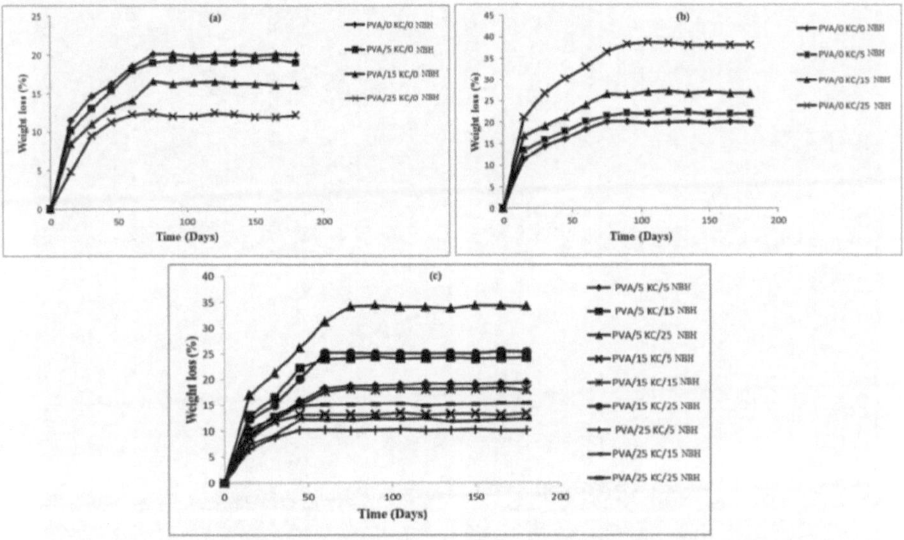

FIGURE 9.5 Weight loss (%) of composite films: (a) PVA/KC films, (b) PVA/NBH films and (c) PVA/KC/NBH films.

higher weight loss compared to unfilled PVA film. As anticipated, NBH was more biodegradable than PVA. The result revealed that the biodegradation rate in terms of weight loss of barley husk-based film increased significantly with the increase in barley husk content during the initial 60 days.

Biodegradation of the films increased with soil burial time and decreased with increased KC content (Figure 9.5c). After 75 days, the PVA film showed 19.02% weight loss which increased to 20.57% after 180 days of soil burial. For 5% NBH in the PVA matrix, the weight loss increased to 22.55% after 75 days of soil burial. However, for 5% KC in hybrid film (PVA/5NBH/5KC), weight loss reduced to 20% after 75 days. The hybrid film containing 25% KC (PVA/5NBH/25KC) showed weight loss of 14.05% over a period of 75 days, which was 6.75% lower than the PVA/5NBH/5KC hybrid film. The highest weight loss of 40% over a period of 180 days was observed for PVA/25NBH/5KC hybrid film.

9.4.5 SEM

The micrographs of poly vinyl alcohol films (Figure 9.6a) were uniform and regular. However, the surface of poly vinyl alcohol films became rough after biodegradation. The results indicated that the film (PVA/5KC/0NBH) had irregular surface and KC particles were also agglomerated on the surface (Figure 9.6e & Figure 9.6f). The micrograph of the PVA/5KC/0NBH composite film appeared to be cracked after biodegradation under soil burial. The micrographs of hybrid film (PVA/5KC/5NBH) were highly irregular and a few holes also appeared on the surface after biodegradation (Figure 9.6g & Figure 9.6h).

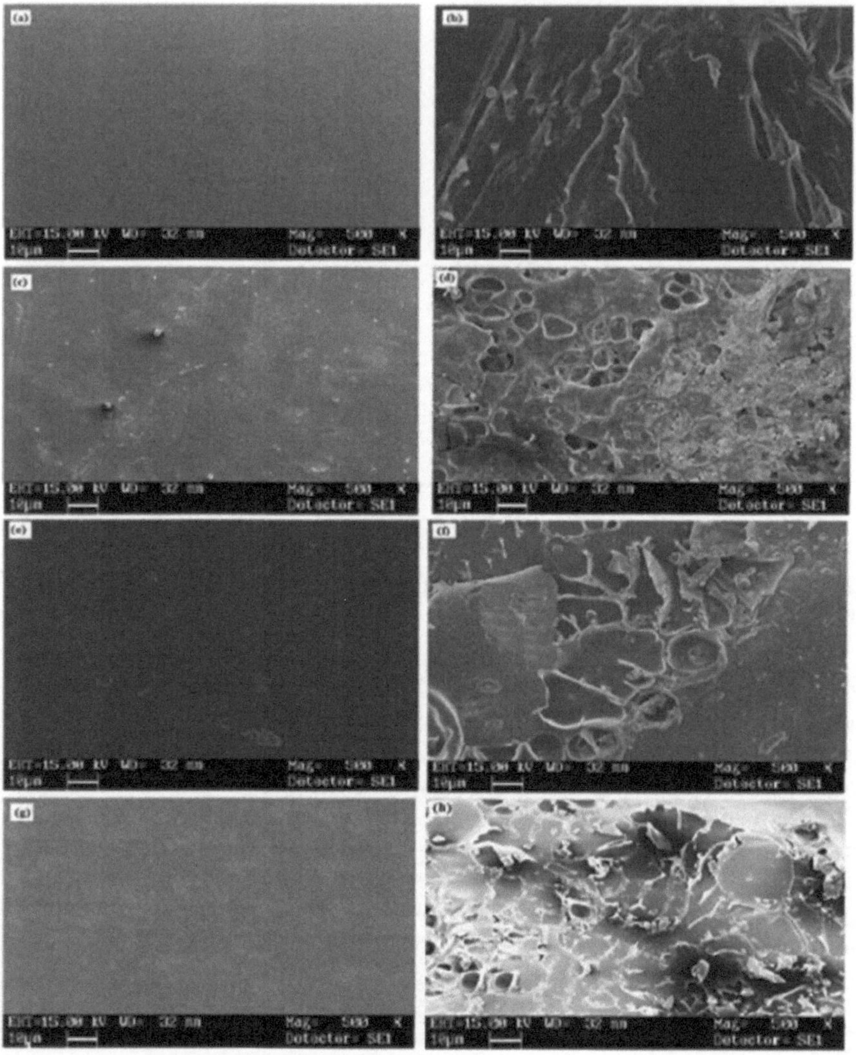

FIGURE 9.6 SEM micrographs before and after biodegradation: (a) neat PVA before, (b) neat PVA film after, (c) PVA/0KC/5NBH film before, (d) PVA/0KC/5NBH film after (e) PVA/5KC/0NBH film before, (f) PVA/5KC/0NBH film after, (g) PVA/5KC/5NBH hybrid film before and (h) PVA/5KC/5NBH hybrid film after.

9.4.6 Comparative Study of Hybrid Films

The comparative analysis of the mechanical properties of hybrid films and others available in the literature is presented in Table 9.2. Hossen et al. (2015) prepared polyethylene hybrid films containing 15% jute fibre and 2% MMT clay and achieved tensile strength of 25.6 MPa and modulus of 1420 MPa. In another study, Islam et al. (2015) reinforced wood (15%), coir (15%) and MMT clay (2 phr) in a polypropylene

TABLE 9.2
Mechanical Properties of Hybrid Films Compared

References	Polymer	Hybrid fillers (w % /w%) or (phr/phr)	Tensile strength (MPa)	Young's modulus (MPa)	Elongation (%)
Hossen et al. (2015)	Polyethylene	Jute fibre/ montmorillonite clay (15%/2%)	25.6	1420	—
Arrakhiz et al. (2013)	Polypropylene	Pine cone fibres/ nanoclay (25%/5%)	27	1330	—
Arjmandi et al. (2015)	Poly (lactic acid)	Cotton linter microcrystalline cellulose/ montmorillonite clay (5 phr/5 phr)	17.5	—	11
Islam et al. (2015)	Polypropylene	Wood/coir/ montmorillonite clay (15%/15%/2phr)	12.1	340	—
Zahedi et al. (2015)	Polypropylene	Almond shell flour/ organically modified montmorillonite clay (50%/5%)	11.6	2700	—
Present study	Poly (vinyl alcohol)	Barley husk/kaolinite clay (5%/5%)	29.33	—	274.03

phr: parts per hundred andw %: weight percentage.

matrix and reported a maximum tensile strength of 12.1 MPa. However, in the present study, hybrid film (poly vinyl alcohol/5% kaolinite clay/5% barley husk) showed tensile strength of 29.33 MPa and elongation of 274%. It was also found that the elongation (%) of poly vinyl alcohol/kaolinite clay/barley husk film was higher than other hybrid films. This was due to the fact that the elongation (%) of unfilled PVA film was 271.57% and it further increased with the increase in KC loading up to 5%. This indicated that the KC increased the plasticity of the films (Tee et al., 2013). This comparative study showed that higher mechanical properties can be achieved using bio fillers and biodegradable polymeric matrix.

9.5 CONCLUSION

Poly vinyl alcohol/barley husk/kaolinite clay-based films were synthesized and their various properties were examined. Films reinforced with barley husk were found to have high water absorption compared to kaolinite clay-reinforced films. Kaolinite clay-based composite films revealed better tensile strength and elongation. SEM micrographs showed that the surface of PVA/NBH films appeared more rough and fragile than those filled with kaolinite clay. Biodegradation studies showed that the weight loss (%) increased with the addition of barley husk. PVA/0KC/25NBH composite film achieved more than 34% weight loss after 60 days. Therefore, these films can be used as a substitute for non-conventional packaging materials.

REFERENCES

Ali, E.S., Ahmad, S. 2012. Bionanocomposite hybrid polyurethane foam reinforced with empty fruit bunch and nanoclay. *Composites Part B: Engineering* 43(7), 2813–2816.

Asaithambi, B., Ganesan, G., Ananda Kumar, S., 2014. Bio-composites: Development and mechanical characterization of banana/sisal fibre reinforced poly lactic acid (PLA) hybrid composites. *Fibers and Polymers*, 15(4), 847–854.

Cheng, Q., Wang, S., Han, Q. 2010. Novel process for isolating fibrils from cellulose fibers by high-intensity ultrasonication. II. Fibril characterization. *Journal of Applied Polymer Science* 115, 2756–2762.

Fortunati, E., Armentano, I., Iannoni, A., Barbale, M., Zaccheo, S., Kenny, J. M. 2012. New multifunctional poly(lactide acid) composites: mechanical, antibacterial, and degradation properties. *Journal of Applied Polymer Science* 124, 87–98.

Fortunati, E., Puglia, D., Luzi, F., Santulli, C., Kenny, J.M., Torre, L. 2013. Binary PVA bio-nanocomposites containing cellulose nanocrystals extracted from different natural sources: part I. *Carbohydrate Polymers* 97(2), 825–836.

Geyer, R., Jambeck, J.R., Law, K.L., 2017. Production, use, and fate of all plastics ever made. *Science Advances*, 3(7), 1–5.

Haq, M., Burgueno, R., Mohanty, A.K., Misra, M. 2008. Hybrid bio-based composites from blends of unsaturated polyester and soybean oil reinforced with nanoclay and natural fibers. *Composites Science and Technology* 68 (15–16), 3344–3351.

Hossain, K.M.K., Ahmed, I., Parsons, A.J., Scotchford, C.A., Walker, G.S., Thielemans W. 2012. Physico-chemical and mechanical properties of nanocomposites prepared using cellulose nanowhiskers and poly(lactic acid). *Journal of Materials Science* 47, 2675–2686.

Hossen, M.F., Hamdan, S., Rahman, M.R., Rahman, M.M., Liew, F.K., Lai, J.C., 2015. Effect of fiber treatment and nanoclay on the tensile properties of jute fiber reinforced polyethylene/clay nanocomposites. *Fibers and Polymers*, 16(2), 479–485.

Ighwela, K. A., Ahmad, A. B., Abol-Munafi, A. B., 2012. Production of Cellulose from Barley Husks as a Partial Ingredient of Formulated Diet for Tilapia Fingerlings. *Journal of Biology, Agriculture and Healthcare*, 2(2), 19–24.

Islam, M.S., Ahmad, M.B., Hasan, M., Aziz, S.A., Jawaid, M., Haafiz, M.K.M., Zakaria, S. A.H., 2015. Natural Fibre-Reinforced Hybrid Polymer Nanocomposites: Effect of Fibre Mixing and Nanoclay on Physical, Mechanical, and Biodegradable Properties. *BioResources*, 10(1), 1394–1407.

Mbey, J.A., Hoppe, S., Thomas, F. 2012. Cassava starch–kaolinite composite film. Effect of clay content and clay modification on film properties. *Carbohydrate Polymers* 88(1), 213–222.

Melo, J.D.D., de Carvalho Costa, T.C., de Medeiros, A.M., Paskocimas, C.A. 2010. Effects of thermal and chemical treatments on physical properties of kaolinite. *Ceramics International* 36(1), 33–38.

Nayak, S.K., Mohanty, S., Samal, S.K., 2009. Influence of short bamboo/glass fiber on the thermal, dynamic mechanical and rheological properties of polypropylene hybrid composites. *Materials Science and Engineering: A* 523 (1–2), 32–38.

Salem, I.A.S., Rozyanty, A.R., Betar, B.O., Adam, T., Mohammed, M., Mohammed, A.M. 2017. Study of the effect of surface treatment of kenaffibre on mechanical properties of kenaf filled unsaturated polyester composite. *Journal of Physics: Conference Series* 908, (1–5).

Sanjay, M.R., Madhu, P., Jawaid, M., Senthamaraikannan, P., Senthil, S., Pradeep, S. 2018a. Characterization and properties of natural fibre polymer composites: a comprehensive review. *Journal of Cleaner Production* 172, 566–581.

Sanjay, M.R., Siengchin, S., Parameswaranpillai, J., Jawaid, M., Pruncu, C. I., Khan, A. 2018b. A comprehensive review of techniques for natural fibres as reinforcement in composites: preparation, processing and characterization. *Carbohydrate Polymers* 207, 108–121.

Tagreed, K.H. 2013. Refractive index dispersion and analysis of the optical parameters of (PMMA/PVA) thin film. *JournalofAlNahrain University Science* 16(3), 164–170.

Tan, B., Ching, Y., Poh, S., Abdullah, L., Gan, S. 2015. A review of natural fibre reinforced poly(vinyl alcohol) based composites: application and opportunity. *Polymers* 7(11), 2205–2222.

Tan, B.K., Ching, Y.C., Gan, S.N., Ramesh, S., Rahman, M.R., 2014. Water absorption properties of kenaffibre–poly(vinyl alcohol) composites. *Materials Research Innovations* 18, 144–146.

Tee, T.T., Sin, L.T., Gobinath, R., Bee, S.T., Hui, D., Rahmat, A.R., Fang, Q. 2013. Investigation of nano-size montmorillonite on enhancing polyvinyl alcohol–starch blends prepared via solution cast approach. *Composites Part B: Engineering*, 47, 238–247.

Zahedi, M., Khanjanzadeh, H., Pirayesh, H., Saadatnia, M.A., 2015. Utilization of natural montmorillonite modified with dimethyl, dehydrogenated tallow quaternary ammonium salt as reinforcement in almond shell flour–polypropylene bio-nanocomposites. *Composites Part B: Engineering*, 71, 143–151.

10 Damage to Polymer Matrix in Transport Applications

T. V. Vineeth Kumar and N. Shanmuga Priya
Siddaganga Institute of Technology, Tumakuru, India

S. Arun
National Aerospace Laboratory, Bengaluru, India

CONTENTS

10.1 INTRODUCTION

Polymers are made up of carbon and hydrogen branches that are chemically joined together to form a chain. Thermoplastic polymers, thermosetting polymers and elastomers are examples of composite polymers. Polymer composite fabrics are made up of a polymer matrix (resin) and a binding agent (fibres). Furthermore, fillers, modifiers and additives can be added to FRP composites to change their properties, boost their efficiency and contribute to cost reduction (Alberto, 2013).

DOI: 10.1201/9781003128861-10

After curing, thermosets have the characteristics of a well bonded three-dimensional molecular structure. They decompose instead of melting as they harden. Thermoplastics are materials with a one- or two-dimensional molecular structure and a high melting point. Spray lay-up, hand lay-up, pultrusion, vacuum moulding, resin transfer moulding (RTM) and filament winding are integrated processes in the fabrication and forming of polymer matrix composites into finished products (Hussain et al., 2006).

From electrical components to a vast variety of vehicle parts, PMCs can be used in nearly every aspect of daily life. The key advantages of PMCs are their low weight, together with their high stiffness and mobility in the direction of the reinforcement. Their utility in aircraft, cars and other moving objects is based on this combination of properties. The material has better corrosion and fatigue resistance than metals. Since the matrix decomposes at high temperatures, contemporary PMCs will only operate at temperatures around 316 °C (Pb-, 1988). The high strength or stiffness-to-weight ratio of PMCs makes them desirable for marine and aerospace applications (Rajak et al., 2019). They are lightweight and have high specific strength while still reducing total weight by 20% to 40% (Ghori et al., 2018a). PMCs have also found application in aerospace due to their dimensional stability, lower thermal expansion properties, and excellent fatigue and fracture resistance. Modern combat fighter aircraft have lost 30% of their weight due to polymer composite materials (Ghori et al., 2018b). However, among the drawbacks higher recurrent and non-recurring costs, non-visible impact damage, higher material costs, lack of low-cost processing methods, flammability, toxicity and a lack of evidence on the impact of long-term exposure to service environments on the properties of PMCs (Zhu et al., 2018).

Around 50% of the transportation industry uses fibre-reinforced plastics (Robinson et al., 2012a). One of the main challenges for structural designers is the need to predict the failure mode of fibre-reinforced composites under complex loading conditions (Mallick, 2007). The use of failure criteria is essential for carrying out design activities without resorting to prolonged and expensive experimental programmes. Hence, for a long time, failure criteria for composites have been receiving a great deal of attention from the composite community. Damage to polymer matrix composites in transportation applications is discussed in this chapter.

10.2 USE OF POLYMER COMPOSITE MATERIALS IN TRANSPORTATION APPLICATIONS

With their many advantages in both use and fabrication processes, composites are found in a remarkably broad variety of contexts. Composite materials are widely used in the transportation (cars, vessels, etc.) and building industries, as well as in chemical corrosion equipment and electric or electronic appliances. Indeed, it is hard to find any industry that does not utilize their benefits. Composite materials are used in the aerospace and marine industries to reduce fuel consumption and also for their anti-corrosion and anti-vibration properties. lIn order to minimize emissions, many countries are also progressively developing the use of natural gas to replace petroleum as a gasoline and diesel fuel. The primary reason for the growth in composite applications is that products made of composites possess a high specific strength-to-weight ratio.

Aerospace applications account for about 50% of current sales of advanced polymer composites, but only about 1–2% of total production. Another 25% of the market is mainly composed of sporting goods like golf clubs and tennis rackets (Chen, 2014). PMCs are used in 25% of automobiles and industrial machinery (Brooks et al., 2017). The next major challenge will be to implement the use of PMCs in massive military and commercial transport aircraft. PMCs make up about 3% of the structural weight of a commercial aircraft like the Boeing 757, but they have the potential to make up over 65% of the overall weight (Koniuszewska and Kaczmar, 2016). The primary reason for using PMCs in commercial aircraft is to save money on diesel. PMCs provide the same structural properties as monolithic metal. In the Airbus A380, the upper fuselage is made from a fibre/metal laminate (FML) called GLARE. It has a lower density and better mechanical properties than monolithic aluminum. Figure 10.1illustrates the popular use of GLARE in the Airbus A380 (Richter-Trummer et al., 2011).

Automobile manufacturers are developing new lightweight materials to improve vehicle safety, noise and vibration, and fuel efficiency. Glass-fibre unsaturated polyester is the most common composite material used in automotive components (Komornicki et al., 2017). The automotive industry, especially car makers, has shown the potential to transition from steel to aluminium alloys, and now there is a quicker transition from aluminium to magnesium and FRP composites, which have better performance. Some experts expect that in the future, the bulk of vehicle weight will be made up of chemical composites lined with fibres (Chauhan et al., 2019).

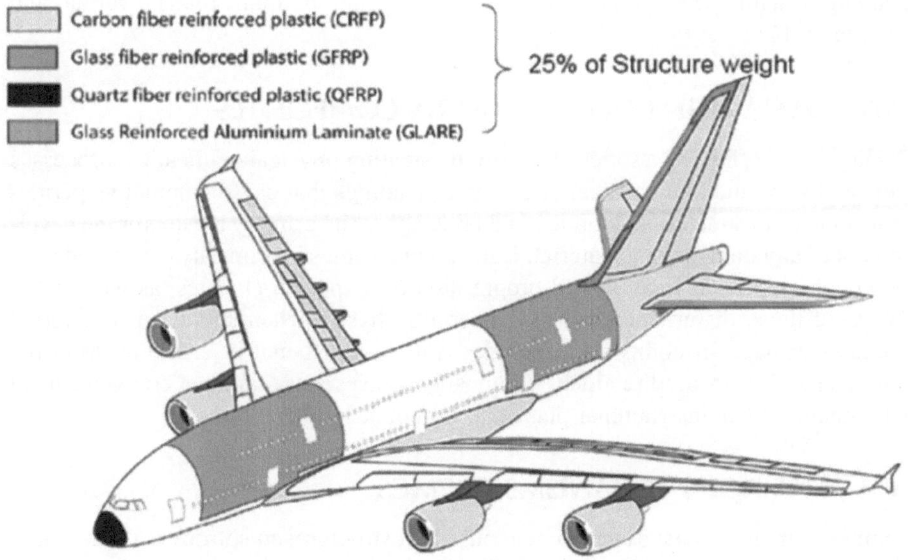

FIGURE 10.1 Airbus A380 structure (Richter-Trummer et al., 2011), image courtesy of Airbus.

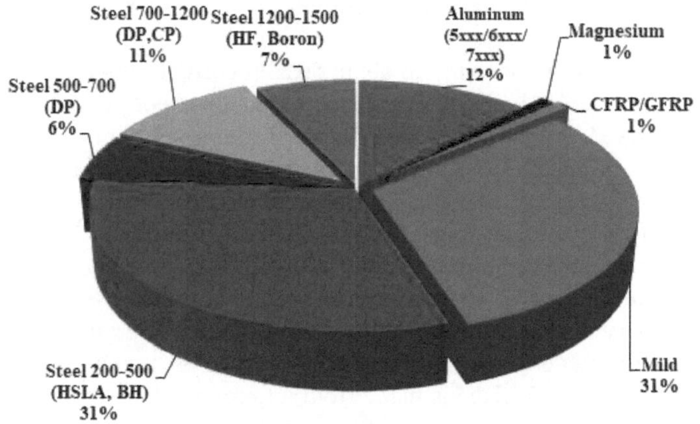

FIGURE 10.2 Contribution of various materials in a typical automobile. Source: CAR Research, 2017.

Magnesium and polymer composites are used in some parts of higher-end cars. Figure 10.2 depicts the use of various materials in a modern car.

Fishing helicopters, tugboats, yachts, boats and minesweepers have also used composite materials in the past. Composite materials are compact and can withstand a lot of external friction. Natural fibre-reinforced composites (NFRP) are being used more and more in the industrial, aerospace, marine and packaging sectors. High water absorption, variations in fibre properties, lack of adhesion between the fibre and the matrix, poor fire resistance and low processing temperature are the main challenges with NFRP composites for transportation applications (Verma and Sharma, 2017).

10.3 DAMAGE IN POLYMER MATRIX COMPOSITES

Damage is described as a series of energy-dissipating physical or chemical processes induced by thermal, mechanical or electrical loadings that cause cumulative permanent changes in a material. Atomic bond breakage is the primary reason for the development of any damage in a material. Damage mechanics is primarily concerned with the conditions for nucleation and progression of dispersed changes, as well as the effects of those disturbances on the material's physics when subjected to external loading. Damage in composites includes fibre–bridged matrix cracking. Multiple matrix cracking in unidirectional composites causes intra-laminar cracking, local delamination in an interlaminar plane and fibre/matrix interfacial slip.

10.4 DAMAGE MECHANISMS IN PMCS

Composite materials have a heterogeneous microstructure; anisotropy is induced due to the presence of boundaries and reinforcement directionality observed in the geometrical features of micro cracks. In addition, the chances of multiple cracking are high when stress transfer takes place via the interfaces between matrix and fibres and plies in a laminate. In the literature, observations on various cracking processes are termed "damage mechanisms".

10.4.1 Interfacial Debonding

The efficiency of an FRP composite is largely determined by the properties of the fibre-matrix interface. The adhesion bond in the interface affects the macroscopic properties. This interface is critical for stress transfer between constituents of the composite materials. Interfacial slipping and fibre pull-out are often influenced by mobility constraints between the fibre and matrix. The shear stress at the interface exceeds the interfacial shear pressure over a long period across the fibre, causing debonding. Debonding initiation and stress transfer at the interface are commonly predicted using shear lag and coherent zone models (Hild and Feillard, 1997).

10.4.2 Matrix Micro Cracking/Intra-laminar Cracking

Fibre-reinforced composites possess superior mechanical properties like strength and stiffness along the fibres, but across the fibres they are generally weak. As a result, they are prone to cracking along fibres. In fibre-reinforced composites, these cracks are the first point of failure. Tensile loading, fatigue loading and thermal cycling are the primary causes of cracks in the longitudinal direction (Nairn, 2000). Debonding of fibres and matrixes, as well as manufacturing flaws such as voids and inclusions, may cause them.

10.4.3 Interfacial Sliding

Interfacial sliding is the differential displacement between the constituents of a composite (Adams, 1987). Matrix crack is the primary source of interfacial sliding; sometimes it can also result from growth of interfacial defects. As a result, debonding-induced interfacial sliding may be called a different mode of damage or a damage mode linked to matrix damage (Womack and Ingber, 2008).

10.4.4 Delamination/Interlaminar Cracking

When two adjacent matrix cracks in a fibre-reinforced composite laminate are joined, an interlaminar delamination crack forms. Delamination occurs in a laminate where the interfacial plane between two adjoining lamina cracks, causing the lamina to separate (Prombut et al., 2006).

10.4.5 Manufacturing Defects

Manufacturing defects in a polymer matrix composite may exist in either the fibres or the matrix. Voids are one of the primary defects that can be observed in almost all types of composite material. The presence of even low-volume fraction voids can cause deleterious effects on flexural, transverse and shear properties of the material (Allemang et al., 2014).

10.4.6 Conclusion

The various damage mechanisms explained above have different characteristics according to geometric and material parameters. The growth of each mechanism depends on the type of load. In short-fibre composites, three basic interfacial failure damage mechanisms have been described (Sirivedin et al., 2000):

1. Mode α: Due to the stress accumulation at the fibre end, localized matrix yielding occurs at the interface.
2. Mode β: An interface crack propagates from the debonded fibre end if the interface is relatively weak.
3. Mode γ: A conical matrix crack propagates from the debonded fibre end at an angle to the fibre axis if the interface is relatively strong.

Similarly for particulate composites, debonding of the particle and cavity nucleation are the major damage mechanisms (Ravichandran and Liu, 1995). Damage dynamics is the field of engineering that studies the causes and repercussions of microstructural events in solids that cause changes in their reaction to external loading. The general goals of damage dynamics research are to identify the causes of the first damage occurrence, forecast the progression of progressive damage, classify and measure structural damage, assess damage criticality and resilience, and provide insights into overall structural analysis and design.

10.5 FAILURE MECHANISMS OF POLYMER MATRIX COMPOSITES IN TRANSPORT APPLICATIONS

It is important to consider potential failure modes, most particularly delamination, and interfacial debonding when designing different structural configurations and manufacturing approaches. To prevent potential damage, temperature variations, moisture and a variety of fluid interactions also need to be taken into account in the selection of materials and the design of structures for transport applications. Understanding the failure mechanisms of composite materials is essential in order to reap their advantages without introducing unintended stress concentrations and/or failure modes that jeopardize structural integrity. The failure modes of laminated composite materials are depicted in Figure 10.3.

10.5.1 Aerospace Applications

Since the development of composite materials, they have been preferred for use in aerospace structures due to their reduced weight compared with metallic structures. Composite joints, especially in aerospace applications, are of particular importance; they act as damage initiation areas as the stress concentration is high in the proximity of these joints. In structures, fasteners and joints are important as they are involved in load transfer from one substructure to another. Hence, the local stress concentration must be considered in the design of joints. Another important failure mechanism in composite material is impact damage, as it differs from impact damage in metals.

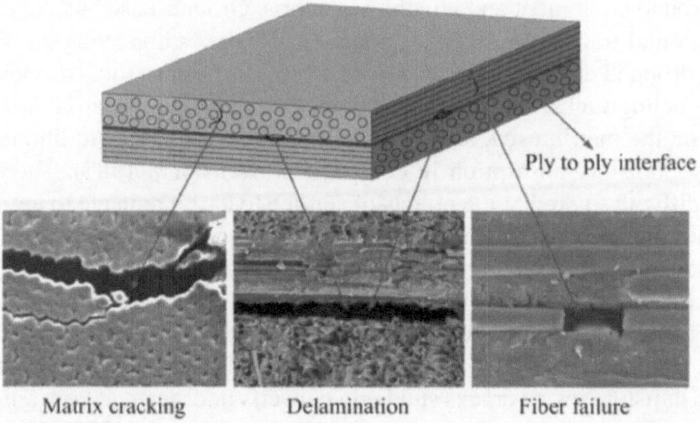

FIGURE 10.3 Failure modes in laminated composite materials (Cao et al. 2020).

In metals the effect of any impact can be visualized in the form of surface dents, whereas composites may not show any significant evidence at all. In the case of untoughened composites, subsurface damage may occur with little or no evidence on the surface of the laminate. Damage test methods have been developed for composites to determine the compressive strength after impact and barely visible impact (Minakuchi et al., 2009). Barely visible impact damage is used to define the critical damage level that establishes the design strength parameters in accordance with the standards.

Along with the basic structural requirements there are many other sources of potential damage related to the in-service environment. These include weather conditions like humidity, heat, cold, wind, rain, ice, snow, dust and lightning, animal interactions (birds, mammals, insects) and ground-based operations (Sun and Hallett, 2017). Impacts during flight are usually bird strikes, ice impingement and airborne debris. Figure 10.4 shows damage to an EgyptAir Boeing 737-700 caused by a deadly bird hit. Ground-based impacts are generally due to debris on runways, direct

FIGURE 10.4 Huge hole at the nose of EgyptAir Boeing 737-700 due to bird strike (Copyright AmirHashim/SWNS.com).

hits from ground equipment and windblown debris. Ground-based damage is mainly caused by ground-traffic control errors, maintenance crews impacting the airframe or tools being dropped during servicing. Ground vehicles such as food-servicing trucks, forklifts, refuelingtrucks, etc. may cause damage to the airframe structure. In FRPs, depending on the matrix used, some moisture will be absorbed. Additionally, stress can increase moisture absorption in composites (Weitsman, 1987). Therefore it is frequently difficult to predict the effects of service loads on damage to aircraft structures, as environmental factors may sometimes be more severe. In the case of a honeycomb, moisture accumulating within the empty space adds to an aircraft's weight. This moisture has the potential to suddenly vaporize, which causes a large pressure rise inside the sandwich structure that may cause the skins to delaminate from the core. In a honeycomb-sandwich structure a phenomenon called "thermal spike" may occur when moisture inside cracks suddenly converts into steam at high temperatures and results in damage within the composite (McKague et al., 1975). In addition to these environmental factors, there are many other considerations during material development and structural design. Since carbon fibres are widely used for aerospace applications, the matrix is susceptible to degradation. The development of more damage-resistant materials and damage-tolerant technologies are two simple and related, ways of enhancing damage tolerance in FRPs. Bolts, pins and adhesives are used to connect aircraft components, and each joint form will behave differently under stress, affecting all materials and structures in adjacent structures on the aircraft.

It is clear that many variables contribute to damage in polymer matrix composite aircraft structures, and it is important to understand the relationship between materials, manufacturing methods and testing. The designer must take note of potential damage modes based on expected damage sources. Since these sources are unlikely to change, damage tolerance of composite materials-based aircraft structures has to improve through development of a combination of materials, design concepts, manufacturing methods and analysis methods.

10.5.2 Marine Applications

FRP composites have been widely used for some decades in marine applications such as radomes, bulk frames, superyachts, working boats and pleasure craft. Bearings, propellers, industrial hatch covers, exhausts and topside systems are a few of the less well-known uses of FRPs. Most of the first significant applications of glass-fibre composites (GRP) were in marine environments (Rudd et al., 1997), pioneering the building of massive composite structures in many industries. In marine structures, damage is principally caused by unexpected loadings such as a rigid impact causing internal structure degradation or hull damage. Shear behavior of the core or weak interfaces between core and skin are common reasons for failure in sandwich structures. But for most damage mechanisms, the main problem arises because of manufacturing processes.

The materials used for marine structures are different from those used for aerospace applications. In FRPs E-glass fibres are the most popular reinforcement, and unsaturated polyester resins are a widely used matrix material, particularly in the pleasure boat industry. Another type of material commonly used in marine structures uses glass/resin facings laminated onto foam made up of PVC or balsa wood cores. For cylindrical underwater structures, filament-wound composites and syntactic foams are used. In filament-wound composites the fibres from roving are impregnated in the resin bath and wound around the mandrill. The synthetic foam is composed of thin-wall hollow glass spheres of 10 to 100 microns diameter in a polymer matrix, mostly epoxy (Avalle et al., 2001).

In marine structures the quasi-static loads in service can be measured, but many failures are due to dynamic loads. Typically loadings can be classified as hard and soft impacts. Hard impacts range from dropped tools on deck to collisions with a floating object. Figure 10.5 is an example of a marine offshore application, the failure of a composite structure used in the connecting arm of a racing yacht. Soft impact damage is uncommon on fast composite craft and is the result of single or multiple pressure pulses on the hull structure (Faltinsen, 2000). Depending on the sandwich/laminate construction, damage may vary from slight matrix cracking to complete perforation, according to the geometry and the impact energy. Assembly failures are vulnerable and in composites they are associated with stress concentration regions. To avoid early debonding, the design of the interface region between panels and stiffeners is critical. In case of composite tubes, damage tolerance is very important. Tests have been carried out on glass and carbon reinforced composite cylinders to test the growth of impact damage and to examine its influence on pressure test behaviour and failure modes. (Gning et al., 2005, Davies et al. 2005). The failure modes of impacted tubes subjected to external pressure testing show premature local implosion. The main damage mechanism in syntactic foams is loss of buoyancy of the material and damage to spheres of foam.

The closed mould-forming methods and traditional materials used in the marine industry are changing, and FRPs are becoming popular for marine structures. For high-performance applications in the offshore industry and for environmental applications in renewable marine energy structures, the material requirements are closer to those of aerospace structures. Long-term durability quality controls and certification tests must be conducted to understand damage and failure mechanisms. Further research and development efforts are necessary to develop the predictive tools required for new applications.

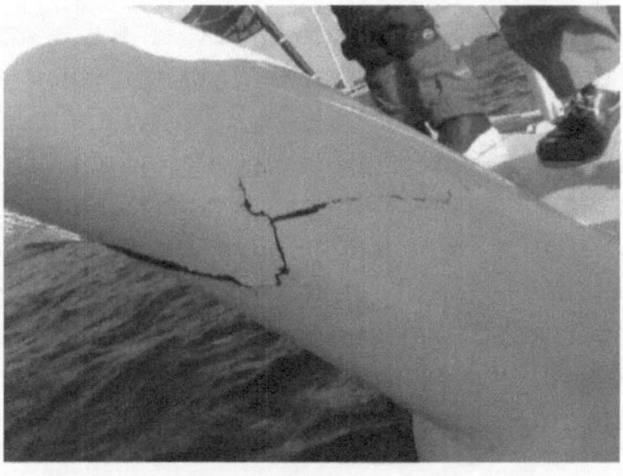

FIGURE 10.5 Composite failure on racing yacht connecting arm (Robinson et al., 2012b).

10.5.3 Automotive Applications

In the automotive industry, polymer matrix composites are widely used to manufacture lightweight cars with improved fuel economy. The most common materials used in the production of automobile parts are sheet-moulding compound (SMC) composites and carbon fibre-reinforced polymer matrix composites. Because of the lower cost, e-glass fibre is used instead of carbon fibre. Epoxy resin matrix composites are not usually used in the automotive industry because they take more time to cure and are more costly than polyesters or vinyl esters. For functional, interior and semi-structural components thermoplastics are used, while body panel, body structure and suspension parts are made from thermosets (Robinson et al., 2012b). Vehicle chassis and body-in-white components are subjected to fatigue loading due to uneven road surface conditions, bumps and potholes. In a high-speed collision, the vehicle body is subjected to multiple impulse loads leading to global buckling, crumpling, tearing and bending failures.

The vehicle's working environment is also important in predicting potential damage areas. Components near the exhaust system or engine, for example, will experience temperatures above 100 °C. In certain regions humidity and moisture changes will affect the failure of critical parts. During the winter season in cold regions vehicles may be exposed to salty road surfaces. Vehicles involved in high-speed accidents may catch fire. Particular attention must be paid to carbon fibre-reinforced polymer composites, as repairable carbon fibres will be released into the atmosphere. Both tensile strength and modulus of SMC composite structures are inversely proportional to temperature (Goh et al., 2018), mainly, mainly because of the softening and weakening of the matrix with rising temperature.

The stress strain diagram for SMCs is somewhat nonlinear below the yield point. This is an indication of damage formation even at low strain levels (Watanabe and Yasuda, 1982). Debonding takes place at the fibre matrix interface. Cracks begin to form in the resin matrix as micro cracks in physical appearance at several locations in the material and propagate till the fibres are interrupted. With increased loading, multiple matrix cracks will be formed at random places in the material, leading to fibre pull-out separation, fibre fracture and fibre–matrix interface debonding. This damage continues to increase as several micro cracks unite together to form a major crack throughout the specimen until complete fracture occurs. In SMC automotive structures, fatigue damage appears as micro cracks in the matrix, and fibre–matrix interface debonding takes place at a lower number of cycles (Laribi et al., 2017). Common low-energy impacts are accidental bumps with another vehicle, hailstorms, stone chip impacts, etc. In low-impact damage the principal micro-failure modes observed are fibre fracture, fibre pull-out, matrix cracking and fibre–matrix interface debonding. In polymer matrix composite tubular structures under quasi-static compressive loading, depending on the fibre/matrix properties, the damage modes identified are fragmentation mode, splaying mode, folding mode and brittle fracture mode (Dayong et al., 2018). Crushing experiments conducted for PMC tubes and energy dissipation during crushing experiments happens through several micro-failure mechanisms.

Fatigue or impact cause structural damage to polymer matrix composites used in automobile and road transport applications. There are no established standards in the automotive industry for assessment of structural damage and also regarding reuse of composite structures. Repair of damaged PMC components is difficult and replacement of damaged structure is not advisable as it is expensive. As in the case of aerospace and marine composite applications, in automotive applications once the vehicles are delivered to customers, there is far less regular inspection of the vehicle or its components. It is difficult, as well as uneconomic, to offer non-destructive inspection methods at regular intervals during the service operation of vehicles. Hence, whenever there is damage to structural composite components either due to fatigue or low-energy impact, it will remain undetected until structural failure occurs.

10.6 CONCLUSION

This chapter has discussed damage mechanisms in PMCs used in three major transport applications: aerospace, marine and automotive. Good specific mechanical properties in various environmental conditions, resistance to corrosion, and ease of assembly, potentially lower cost and light weight are the main characteristics of PMCs for transport applications. The following failure mechanisms are identified in damage to polymer matrix composites in aerospace, marine and automotive structures: interfacial debonding; matrix micro cracking/intra-laminar cracking; interfacial cracking; delamination/interlaminar cracking; and manufacturing defects.

Although repair and recycling of the PMCs used in transport applications in recent decades is not yet well established, progress is being made in research and development of these aspects.

REFERENCES

Adams, D.F. 1987. Engineering composite materials. *Composites* 18(3): 261. doi:10.1016/0010-4361(87)90420-4.

Alberto, M. 2013. Introduction of fibre-reinforced polymers – polymers and composites: concepts, properties and processes. *Fiber Reinforced Polymers - The Technology Applied for Concrete Repair*, 3–40. doi:10.5772/54629.

Allemang, R., J. De Clerck, C. Niezrecki, and A. Wicks. 2014. Preface. *Conference Proceedings of the Society for Experimental Mechanics Series* 45(7): 577–617.

Avalle, M., G. Belingardi, and R. Montanini. 2001. Characterization of polymeric structural foams under compressive impact loading by means of energy-absorption diagram. *International Journal of Impact Engineering*. doi:10.1016/S0734-743X(00)00060-9.

Brooks, R., S.M. Shanmuga Ramanan, and S. Arun. 2017. Composites in automotive applications: design. *Reference Module in Materials Science and Materials Engineering*. doi:10.1016/b978-0-12-803581-8.03961-8.

Cao, Y., Z. Cao, Y. Zhao, D. Zuo, and T.E. Tay. 2020. Damage progression and failure of single-lap thin-ply laminated composite bolted joints under quasi-static loading. *International Journal of Mechanical Sciences* 170: 105360. doi:10.1016/j.ijmecsci.2019.105360.

Chauhan, V., T. Kärki, and J. Varis. 2019. Review of natural fiber-reinforced engineering plastic composites, their applications in the transportation sector and processing techniques. *Journal of Thermoplastic Composite Materials*. doi:10.1177/0892705719889095.

Chen, Y. Y. 2014. The application of carbon fiber materials in sports equipment. *Applied Mechanics and Materials* 687–691 (Emim): 4240–4243. doi:10.4028/www.scientific.net/AMM.687-691.4240.

Davies, P., L. Riou, F. Mazeas, and P. Warnier. 2005. Thermoplastic composite cylinders for underwater applications. *Journal of Thermoplastic Composite Materials*. doi:10.1177/0892705705054397.

Dayong, H., Y. Wang, L. Dang, and Q. Pan. 2018. Energy absorption characteristics of composite tubes with different fibers and matrix under axial quasi-static and impact crushing conditions. *Journal of Mechanical Science and Technology* 32: 2587–2599. doi:10.1007/s12206-018-0516-y.

Faltinsen, O.M. 2000. Hydroelastic slamming. *Journal of Marine Science and Technology* 5(2): 49–65. doi:10.1007/s007730070011.

Ghori, S. W., R. Siakeng, M. Rasheed, N. Saba, and M. Jawaid. 2018a. The role of advanced polymer materials in aerospace. *Sustainable Composites for Aerospace Applications* 6282038: 19–34. doi:10.1016/B978-0-08-102131-6.00002-5.

Ghori, S.W., R. Siakeng, M. Rasheed, N. Saba, and M. Jawaid. 2018b. The role of advanced polymer materials in aerospace. *Sustainable Composites for Aerospace Applications*. Elsevier Ltd. doi:10.1016/B978-0-08-102131-6.00002-5.

Gning, P. B., M. Tarfaoui, F. Collombet, L. Riou, and P. Davies. 2005. Damage development in thick composite tubes under impact loading and influence on implosion pressure: experimental observations. *Composites Part B: Engineering*. doi:10.1016/j.compositesb.2004.11.004.

Goh, G.D., V. Dikshit, A.P. Nagalingam, G.L. Goh, S. Agarwala, S.L. Sing, J. Wei, and W.Y. Yeong. 2018. Characterization of mechanical properties and fracture mode of additively manufactured carbon fiber and glass fiber reinforced thermoplastics. *Materials and Design*. doi:10.1016/j.matdes.2017.10.021.

Hild, F. and P. Feillard. 1997. Ultimate strength properties of fiber-reinforced composites. *Reliability Engineering & System Safety* 56(3): 225–235. doi:10.1016/S0951-8320(97)00094-X.

Hussain, F., M. Hojjati, M. Okamoto, and R.E. Gorga. 2006. Review article: Polymer-matrix nanocomposites, processing, manufacturing, and application: an overview. *Journal of Composite Materials* 40 (17): 1511–1575. doi:10.1177/0021998306067321.

The ASHA Leader. 2017, June. *The ASHA Leader* 22 (6): 20–22. doi:10.1044/leader.ppl.22062017.20.

Komornicki, J., L. Bax, H. Vasiliadis, I. Magallon, and K. Ong. 2017. *Polymer composites for automotive sustainability. SusChem, European Technology Platfrom for Sustainable Chemistry, Brussels.* http://www.suschem.org/files/library/Publications/POLYMERS_Brochure_Web.pdf

Koniuszewska, A.G., and J.W. Kaczmar. 2016. Application of polymer based composite materials in transportation. *Progress in Rubber, Plastics and Recycling Technology* 32(1): 1–23. doi:10.1177/147776061603200101.

Laribi, M., S. Tamboura, J. Fitoussi, R. Tiébi, A. Tcharkhtchi, and H. Dali. 2017. Fast fatigue life prediction of short fiber reinforced composites using a new hybrid damage approach: application to SMC. *Composites Part B: Engineering* 139 (December). doi:10.1016/j.compositesb.2017.11.063.

Mallick, P.K. 2007. *Fiber-Reinforced Composites: Materials, Manufacturing, and Design.* CRC Press. Boca Raton, FL. https://doi.org/10.1201/9781420005981.

McKague, E.L., J.E. Halkias, and J.D. Reynolds. 1975. Moisture in composites: the effect of supersonic service on diffusion. *Journal of Composite Materials*. doi:10.1177/002199837500900101.

Minakuchi, S., Y. Okabe, T. Mizutani, and N. Takeda. 2009. Barely visible impact damage detection for composite sandwich structures by optical-fiber-based distributed strain measurement. *Smart Materials and Structures*. doi:10.1088/0964-1726/18/8/085018.

Nairn, J. A. 2000. Matrix microcracking in composites. *Comprehensive Composite Materials* 2: 403–432. doi:10.1016/b0-08-042993-9/00069-3.

Pb-, N. 1988. Advanced materials by design june 1988. *Advanced Materials*.

Prombut, P., L. Michel, F. Lachaud, and J. Barrau. 2006. Delamination of multidirectional composite laminates at 0°/[Theta]° ply interfaces. *Engineering Fracture Mechanics* 73: 2427–2442. doi:10.1016/j.engfracmech.2006.05.013.

Rajak, D.K., D.D. Pagar, P.L. Menezes, and E. Linul. 2019. Fiber-reinforced polymer composites: manufacturing, properties, and applications. *Polymers* 11 (10). doi:10.3390/polym11101667.

Ravichandran, G. and C.T. Liu. 1995. Modeling constitutive behavior of particulate composites undergoing damage. *International Journal of Solids and Structures*. doi:10.1016/0020-7683(94)00172-S.

Richter-Trummer, V., P.M.G.P. Moreira, and P.M.S.T. de Castro. 2011. *Damage tolerance of aircraft panels taking into account residual stress*, 173–194. doi:10.1007/8611_2011_55.

Robinson, P., E. Greenhalgh, and S. Pinho. 2012a. *Failure Mechanisms in Polymer Matrix Composites: Criteria, Testing and Industrial Applications*. doi:10.1533/9780857095329.

Robinson, P., E. Greenhalgh, and S. Pinho. 2012b. *Failure Mechanisms in Polymer Matrix Composites*. doi:10.1533/9780857095329.

Rudd, C.D., A.C. Long, K.N. Kendall, and C.G.E. Mangin. 1997. Introduction to liquid composite moulding. In *Liquid Moulding Technologies*. doi:10.1533/9781845695446.1.

Sirivedin, S., D.N. Fenner, R.B. Nath, and C. Galiotis. 2000. Matrix crack propagation criteria for model short-carbon fibre/epoxy composites. *Composites Science and Technology* 60: 2835–2847.

Sun, X. and S. Hallett. 2017. Barely visible impact damage in scaled composite laminates: experiments and numerical simulations. *International Journal of Impact Engineering* 109 (June). doi:10.1016/j.ijimpeng.2017.06.008.

Verma, D. and S. Sharma. 2017. Green biocomposites: a prospective utilization in automobile industry. In *Green Energy and Technology*. doi:10.1007/978-3-319-49382-4_8.

Watanabe, T. and M. Yasuda. 1982. Fracture behaviour of sheet moulding compounds. Part 1. Under tensile load. *Composites*. doi:10.1016/0010-4361(82)90171-9.

Weitsman, Y. 1987. Stress assisted diffusion in elastic and viscoelastic materials. *Journal of the Mechanics and Physics of Solids*. doi:10.1016/0022-5096(87)90029-9.

Womack, S., and M. Ingber. 2008. "*Interfacial debonding and damage progression in particle-reinforced composites*." 127–135. doi:10.2495/BE080131.

Zhu, L., N. Li, and P.R.N. Childs. 2018. Light-weighting in aerospace component and system design. *Propulsion and Power Research* 7(2): 103–119. doi:10.1016/j.jppr.2018.04.001.

11 A Review of Natural Fibre Composites for Orthopaedic Plate Applications

A. Soundhar
Sri Venkateswara College of Engineering (SVCE), Chennai, India

Murugan Rajesh, V. Lakshmi Narayanan, Anita Jessie, Thulasidhas Dhilipkumar, and K. Jayakrishna
Vellore Institute of Technology, Vellore, India

CONTENTS

11.1 INTRODUCTION

Materials that consist of a blend or combination of two or more micro or macro-elements that vary in form and chemical structure and are essentially insoluble in each other are called composite materials (Soundhar and Jayakrishna 2019). Composites involve two phases: the matrix phase (continuous phase) and the reinforcement phase (discontinuous phase). The matrix phase determines the structure of the composite material, while the reinforcement phase determines its strength. Based on the matrix and reinforcement materials, composites are classified into five types: metal matrix composites, ceramics matrix composites, polymer matrix composites, inter-metallic composites and nanocomposites (Yang et al. 2012). In recent years, the use of polymer matrix composites has increased dramatically across the world. Most researchers prefer polymer composite materials over monolithic materials as they possess a high stiffness-to-weight ratio and better mechanical, thermal and electrical properties

DOI: 10.1201/9781003128861-11

(Dixit et al. 2017). Polymers are categorized into two major groups: thermoplastic and thermoset monomers. Thermoplastics are polymers that soften when heated, permitting moulding, and solidify again as they are cooled. Thermoplastic materials can be remoulded and reused without affecting their physical properties. Unlike thermosetting plastics, thermoplastics can be moulded only at their glass transition temperature (Heckele and Schomburg 2003). Examples of thermoplastics are polystyrene (PS), polyester (PE), polyamide (PA), polycarbonate (PC), polyether sulfone (PES), polypropylene (PP), etc. This chapter discusses the importance of natural fibre-reinforced polymer composites for orthopaedic plate applications.

The femur is the largest and strongest bone in the body and has a good blood supply. A large or high-impact force is needed to break this bone. The average adult male femur is 48 cm in length and 2.34 cm in diameter and can support up to 30 times the weight of an adult (Wolinsky et al. 2001). A femoral fracture is a fracture of the femur (thigh bone). A femoral shaft fracture is defined as a fracture of the diaphysis occurring between 5 cm distal to the lesser trochanter and 5 cm proximal to the adductor tubercle, occurring as a result of chronic, repetitive activity (Khan et al. 2017). Femoral shaft fractures in young people are frequently due to some type of high-energy collision. The most common cause of femoral shaft fracture is a motor vehicle or motorcycle crash. Being hit by a vehicle while walking is another common cause. A lower-force incident, such as a fall from standing, may cause a femoral shaft fracture in an older person with weaker bones. The most common types of femoral shaft fractures include transverse, oblique, spiral, and comminuted and open fractures. Femur bone fracture is also known as Vancouver B1 fracture, to which 75% of periprosthetic fractures are attributed. Typical femur fractures are shown in Figure 11.1.

According to the American Academy of Orthopaedic Surgeons, recovery from a fracture of the femur takes from three to six months. This long healing time creates

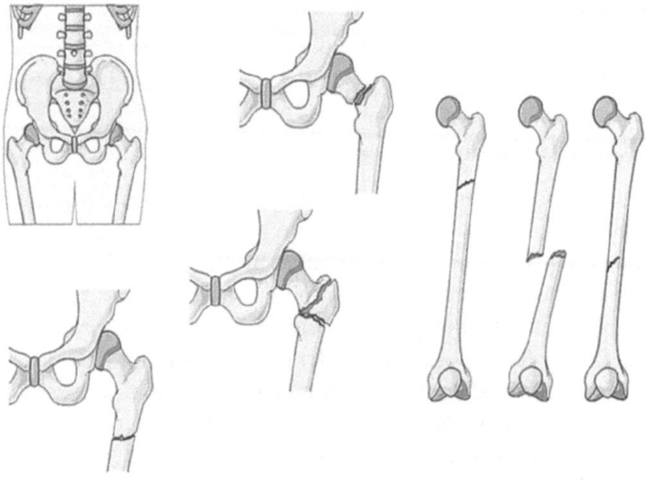

FIGURE 11.1 Types of femur fracture. (Courtesy of https://coreem.net/core/pediatric-femur-fractures/)

the possibility of misalignment (Bartel and Davy, 2006). An orthopaedic plate is thus required to enhance the healing process by arresting the fracture and reducing the gap, healing the fractured bone and supporting proper alignment. The plate reduces the tensile stresses at the fracture site (Scholz et al., 2011). High-rigidity titanium and stainless-steel alloys are used for orthopaedics due to their mechanical stability, bio-inertness, corrosion resistance, and inexpensiveness compared to other biomaterials. However, there is a stiffness imbalance between the human femur and the metal plate, generally called the "stress shielding effect" (Ganesh et al., 2005), which causes more load transition on the plate than on the femur. This disturbs the vascularity of the bone under the plate, inducing bone resorption and decreasing its strength over an extended period. Thus, it is vital to curtail the destructive effect of stress shielding by designing a fracture plate that matches the mechanical properties of the human cortical bone. This is achieved by using polymer-based composite materials as an alternative to metals and alloy-based orthopaedic implants (Rezwan et al., 2006). Moreover, in CT and MRI scanning of the fracture, metallic plates appear white, restricting visibility of the bone, which also appears white. Polymer composite implants, on the other hand, are radiolucent (appearing dark coloured) and are compatible with MRI and CT scans, making the clinical monitoring process easier (Teoh, 2000).

11.2 FIBRE-REINFORCED POLYMER COMPOSITES

Fibre-reinforced polymer composites (FRPC) are now replacing ceramic and metallic materials such as silicon carbide, titanium, stainless steel, cobalt and zirconium.

Polymeric materials currently employed in the field of medicine include polylactic acid (PLA), polycaprolactone, polyethylene glycol (PEG), polyolefin, polyurethanes, silicone, polyamides and polyesters (Jiang et al., 2005; Gollwitzer et al., 2003). Polymeric materials for medical applications are chosen for their engineering properties such as stiffness and strength, and other related properties such as toxicity and biocompatibility (Nair and Laurencin, 2007). The special modifications to these materials required for medical applications are based on their blends, molecular weight, crystallization, crosslinking degree, copolymers and further bioactive surface functionalization (Hamad et al., 2015).

In orthopaedic applications, polymers such as polylactic acid (PLA), polyhydroxyl butyrate (PHB), poly ether ether ketone (PEEK) and polyglycolic acid (PGA) are employed (Daniels et al., 1990; Dhandayuthapani et al., 2011). The major problems with the use of polymers for orthopaedic plates, which insufficient mechanical and wear properties are, can be eliminated by incorporating synthetic fibres into the polymer matrix (Polizu et al., 2006).

Suner et al. (2014) investigated multi-walled carbon nanotube (MWCNT) and the conventional ultra-high molecular-weight polyethylene (UHMWPE). The study compared the volume and size distributions, biocompatibility, wear rate and bioactivity of the wear debris formed from both composites. The wear rate of the composite was considerably lessened by reinforcing MWCNT. Nanocomposites produced with UHMWPE/MWCNT offer better results and are therefore a suitable alternative for orthopaedic plate applications. Robert and Campbell (2006) fabricated carbon fibre-reinforced polymer (CFRP) composite using the inflatable bladder moulding method for femoral stem applications. Improvement in bone density and reduction in fracture susceptibility were found in the CFRP composites. Samiezadeh et al. (2014) produced hybrid composites (carbon fibres/flax/epoxy) for intramedullary nail (IM) applications based on the selective stress shield approach. The introduction of carbon fibre reinforcement in the matrix improved the mechanical and tribological properties of the composite for orthopaedic applications. However, it causes a number of issues, including inadequacies in biocompatibility, and biodegradability. The presence of carbon fibres in the PEEK matrix leads to formation of fragments due to the wear and friction between the implant and bone (Boudeau et al., 2012). It is vital, therefore, to develop light-weighted, cost-effective and biocompatible composites for orthopaedic applications.

11.3 NATURAL FIBRE POLYMER COMPOSITES

Nowadays, natural fibres (pineapple, jute, sisal and hemp) are employed to develop biodegradable composites for orthopaedic applications (Mohammed et al., 2015). Natural fibres are a renewable source in composite fabrication (Faruk et al., 2014) and overcome the undesirable impacts of synthetic fibre polymer composites. Their excellent properties include high stiffness-to-weight ratio, low density, strength-to-weight ratio, and low cost (Gholampour and Ozbakkaloglu, 2020). Table 11.1 categorizes natural fibres according to their plant, mineral and animal sources.

Numerous naturally obtainable materials (flax, *prosopis juliflora*, sisal, betel nut, sugar palm and tamarind fruit fibre) are used as reinforcements in polymer composites to improve environmental sustainability (Unterweger et al., 2014). Recent research seeks potential plant fibres for low- and medium-load transport applications by characterizing their mechanical properties. Woven natural fibres offer favourable properties for broad applications in biomedicine (Cheung et al., 2009). Rao and Rao (2007) reported a study of the tensile properties of bamboo, date and vakka natural fibre-reinforced composites. Compared to bamboo and date fibre composites, excellent tensile properties were found in the vakka fibre-reinforced composites. Murali and Panneerselvam (2011) studied various natural fibre (sisal, coir, kenaf, jute and

TABLE 11.1
Natural Fibre Classification

	Leaf	Pineapple, Banana, Abaca, Sisal
	Bast	Flax, hemp, jute, ramie, kenaf
	Fruit	Coir
	Wood	Softwood, hardwood
	Seed	Cotton, kapok
	Grass/reed	Bamboo, corn
Lignocellulose/Cellulose	Stalk	Wheat, maize, oat, rice
Mineral	—	Asbestos, metal fibres, ceramic fibres
Animal	Hair silk/wool	Cashmere, horsehair, lambswool, goat hair

Sources: Sarikiya et al. (2019); Loganathan et al. (2020).

hemp) reinforced polypropylene composites to assess the mechanical properties. Sisal fibre-reinforced composites yielded higher properties equivalent to those of glass-fibre composite.

Venkateshwaran et al. (2013) investigated banana- and sisal-reinforced composites. Composites offered higher mechanical properties (50 wt.%) and reduced water absorption properties. Li et al. (2007) found that sisal fibre provided better strength, low density and fewer health hazards than other natural fibres. Idicula et al. (2005) studied the characteristics of hybrid banana/sisal composites. Composites with a volume fraction of fibre (0.4%) content provided high flexural and tensile strengths. Sisal-reinforced polyester composite yielded higher impact and damping values than banana-fibre composites. Pothan et al. (2008) investigaged the mechanical properties of woven sisal fibre composite. It was evident that plain 13 woven fibre enhances composite properties. Shibata et al. (2008) examined the flexural strength of random and unidirectionally oriented kenaf/bamboo fibre-reinforced composites. They concluded that woven-fibre composites provided improved flexural modulus and strength. Alavudeen et al. (2015) examined the properties of woven-fibre (banana/kenaf) epoxy composite and randomly oriented short-fibre composites. The woven-fibre composite was found to have superior mechanical properties to the short and randomly distributed fibres.

Xie et al. (2020) studied wear and friction behaviour of UHMWPE composites with sisal fibre filler. Stabilization of frictional properties and improved water absorption was found with sisal fibre fillers in the composites. Reduction of wear debris and wear, and improvement in loading capacity were achieved using the correct percentage of sisal fibre filler. The best anti-friction properties were observed at low speed and heavy load when 10% sisal fibre was added to UHMWPE. This work references the design of sisal fibre-reinforced composites using polymer resin to lessen wear and enhance lubrication.

Senthilkumar et al. (2018) investigated the mechanical properties of sisal/cotton fibre hybrid polyester composite. Results suggest that the combination of natural fibre with another natural or synthetic fibre as reinforcement to the polymer matrix yields enhanced mechanical properties. On the other hand, natural fibres possess certain limitations, such as moisture absorption and hydrophilicity. The common interactions between the surface of the orthopaedic plate, tissue, human cells and other biological fluids increases the wettability and damages the surface properties of the material used for the plate. It is recommended that plates, tissue engineering substrates and devices that come into contact with blood should possess both hydrophilic and hydrophobic characteristics for improved biocompatibility (Yang et al., 2018).

11.4 HYBRID COMPOSITES

Hybrid composites are a combination of natural and synthetic fibre reinforcements of the polymer matrix. Hybrid composites reduce the limitations of natural fibres (Soundhar et al., 2020). John and Naidu (2004) reported a study of hybrid composites with sisal/glass fibre. Higher impact resistance was noticed in the hybrid composite that in the pure sisal fibre composite. Khanam et al. (2010) fabricated a sisal/carbon fibre reinforced hybrid composite. They concluded that the addition of carbon fibre to the matrix improves the tensile and flexural properties. Valente et al. (2011) discussed the effects of adding glass fibre and wood flour to the thermoplastic matrix on its flexural properties. Adding the glass fibre improved the properties of the hybrid composite. Fiore et al. (2017) studied a woven jute and basalt fibre hybrid composite, which showed an improvement in deformation resistance.

Ahmed and Vijayarangan (2008) fabricated woven glass-fibre and woven jute-fibre hybrid sandwich composites to examine their flexural and tensile strength. Glass fibre as an outer layer was found to improve the mechanical properties of the hybrid composite. Similar results with glass-fibre/sisal-fibre hybrid composites were reported by Jarukumjorn and Suppakarn (2009). Gouda et al. (2014) made composites with natural fibres such as sisal, jute, banana, hemp and epoxy reinforcement. Two different volume fractions of fibres (16% and 24%) were used in the study. The physical properties of the composites were compared with those of the femur. The hybrid composites of 24 wt.% fibres yield better mechanical properties than those of other fibre volume fractions. Manteghi et al. (2017) examined a flax/glass fibre/epoxy hybrid sandwich composite used for plates in bone fractures. Bagheri et al. (2015) studied flax/CF/epoxy hybrid composites for orthopaedic trauma applications. There is comparable cell viability with no hazards on gene expression levels between the flax/CF/epoxy composite plate and medical-grade stainless steel. The study suggested the synthetic and natural fibre combination is a suitable replacement for orthopaedic plate applications.

Qiao et al. (2019) developed a nano hydroxyapatite/polyamide66/glass-fibre (n-HA/PA66/glass fibre) composite for the orthopaedic applications. Fractures were fixed by n-HA/PA66/glass fibre plate without any breakages or radiographic imaging intrusion. New bone was also shown to grow at the n-HA/PA66/glass-fibre interface and joined with native bone tissue. Sarwar et al. (2020) produced a new kevlar/flax/epoxy (KFE) sandwich hybrid composite for the orthopaedic plate applications. The KFE sandwich composite enhanced the tension, compression, torsion strength, stiffness and moisture absorption properties. KFE composites offered higher tensile strength and tensile modulus than human cortical bone.

11.5 SUMMARY

Bone implant materials are in huge demand for healing both fractures and deformities. Metal plates may cause osteoporosis, stress shielding and fatigue breakage after prolonged implantation in the body. This study shows the feasibility of cost-effective natural fibre-reinforced polymer sandwich composites as an alternative to orthopaedic plates. Metallic and ceramic prostheses can cause serious problems, such as metal incompatibility with host tissue, anode–cathode reactions, corrosion, magnetism and stress shielding effects and increase in bone porosity (osteoporosis) and maximize the likehihood of secondary sources surgery. Hybrid synthetic/natural fibre-reinforced sandwich polymer composites can be used to reduce the stress shielding effect. Excellent mechanical properties (tensile, flexural, compressive and fatigue) and biomimetic mineralization of synthetic/natural fibre-reinforced sandwich polymer composites can be used to fix a Vancouver B1-type fracture. The benefits of natural fibre-reinforced composite plates are not limited to their unique mechanical properties. The final price of the proposed composite plate is expected to be much cheaper than a clinically used metal one. This provides an additional motivation for pursuing further investigations into the development of natural fibre-reinforced polymer composites as a replacement for metallic orthopaedic plates.

REFERENCES

Ahmed, K. S. and Vijayarangan, S. (2008), 'Tensile, flexural and interlaminar shear properties of woven jute and jute-glass fabric reinforced polyester composites', *Journal of Materials Processing Technology*, **207** (1–3), 330–335.

Alavudeen, A., Rajini, N., Karthikeyan, S., Thiruchitrambalam, M. and Venkateshwaren, N. (2015), 'Mechanical properties of banana/kenaf fiber-reinforced hybrid polyester composites: effect of woven fabric and random orientation', *Materials and Design*, **66**, 246–257.

Bagheri, Z. S., Giles, E., El Sawi, I., Amleh, A., Schematic, E. H. and Zdero, R. (2015), 'Osteogenesis and cytotoxicity of a new carbon fiber/flax/epoxy composite material for bone fracture plate applications', *Materials Science and Engineering: C*, **46** (2015), 435–442.

Bartel, D.L. and Davy, D.T. (2006), *Orthopaedic biomechanics: mechanics and design in musculoskeletal systems*, University of Michigan: Prentice Hall.

Boudeau, N., Liksonov, D., Barriere, T., Maslov, L. and Gelin, J.C. (2012), 'Composite based on polyetheretherketone reinforced with carbon fibres, an alternative to conventional materials for femoral implant: manufacturing process and resulting structural behaviour', *Materials & Design*, **40**, 148–156.

Chen, Q. and Thouas, G.A. (2015), 'Metallic implant biomaterials', *Materials Science and Engineering: R: Reports*, **87**, 1–57.

Cheung, H., Ho, M., Lau, K., Cardona, F. and Hui, D. (2009), 'Natural fibre-reinforced composites for bioengineering and environmental engineering applications', *Composites Part B: Engineering*, **40** (7), 655–663.

Daniels, A. U., Chang, M. K., Andriano, K. P. and Heller, J. (1990), 'Mechanical properties of biodegradable polymers and composites proposed for internal fixation of bone', *Journal of Applied Biomaterials*, **1** (1), 57–78.

Dhandayuthapani, B., Yoshida, Y., Maekawa, T., Kumar, D.S. (2011), 'Polymeric scaffolds in tissue engineering application: a review', *International Journal of Polymer Science*, **2011**, 290602.

Diaz, R., Campbell, D., Diaz Robert, L. and Campbell David, R. (2006), 'Prefracture spinal implant for osteoporotic unfractured bone', U.S. Patent Application 10/976,192.

Dixit, S., Goel, R., Dubey, A., Shivhare, P.R. and Bhalavi, T. (2017), 'Natural fibre reinforced polymer composite materials—a review', *Polymers from Renewable Resources*, **8** (2), 71–78.

Ganesh, V.K., Ramakrishna, K. and Ghista, D.N. (2005), 'Biomechanics of bone-fracture fixation by stiffness-graded plates in comparison with stainless-steel plates', *Biomedical Engineering Online*, **4** (1), 1–15.

Gollwitzer, H., Ibrahim, K., Meyer, H. and Mittelmeier, W. (2003), 'Antibacterial poly (D, L-lactic acid) coating of medical implants using a biodegradable drug delivery technology', *Journal of Antimicrobial Chemotherapy*, **51** (3), 585–591.

Gouda, D. A. T., Jagadish, S. P., Dinesh, K. R. and Gouda, H. (2014a), 'Characterization and investigation of mechanical properties of hybrid natural fiber polymer composite materials used as orthopaedic implants for femur bone prosthesis', *IOSR Journal of Mechanical and Civil Engineering*, **11** (4), 40–52.

Gouda, D. A. T., Jagadish, S. P., Dinesh, K. R. and Gouda, H. (2014b), 'Characterization and investigation of mechanical properties of hybrid natural fiber polymer composite materials used as orthopaedic implants for femur bone prosthesis', *IOSR Journal of Mechanical and Civil Engineering*, 11 (4), 40–52.

Hamad, K., Kaseem, M., Yang, H.W. and Deri, F. (2015), 'Properties and medical applications of polylactic acid: a review', *Express Polymer Letters*, **9** (5), 435–455.

Heckele, M. and Schomburg, W. K. (2003), 'Review on micro molding of thermoplastic polymers', *Journal of Micromechanics and Micro Engineering*, **14** (3), R1.

Huang, Z. M., and Fujihara, K. (2005), 'Stiffness and Strength design of composite bone plates', *Composite Science and Technology*, **65** (1), 73–85.

Idicula, M., Malhotra, S. K., Joseph, K., and Thomas, S. (2005), 'Dynamic mechanical analysis of randomly oriented intimately mixed short banana/sisal hybrid fibre reinforced polyester composites', *Composites Science and Technology*, **65** (7–8), 1077–1087.

Jarukumjorn, K. and Suppakarn, N. (2009), 'Effect of glass fiber hybridization on properties of sisal fiber-polypropylene composites', *Composites Part B: Engineering*, **40** (7), 623–627.

Jiang, G., Evans, M. E., Jones, I. A. and Rudd, C.D. (2005), 'Preparation of poly (epsilon-caprolactone)/continuous bioglass fibre composite using monomer transfer moulding for bone implant', *Biomaterials*, **26** (15), 2281–2288.

Li, X., Tabil, L. G. and Panigrahi, S. (2007), 'Chemical treatments of natural fiber for use in natural fiber-reinforced composites: a review', *Journal of Polymers and the Environment*, **15** (1), 25–33.

Loganathan, T. M., Sultan, M. T. H., Jawaid, M., Shah, A. U. M., Ahsan, Q., Mariapan, M. and Abdul Majid, M. S. (2020), 'Physical, thermal and mechanical properties of areca fibre reinforced polymer composites—an overview', *Journal of Bionic Engineering*, **17** (1), 185–205.

Manteghi, S., Mahboob, Z., Fawaz, Z. and Bougherara, H. (2017), 'Investigation of the mechanical properties and failure modes of hybrid natural fiber composites for potential bone fracture fixation plates', *Journal of the Mechanical Behavior of Biomedical Materials*, **65**, 306–316.

Mohammed, L., Ansari, M. N., Pua, G., Jawaid, M. and Islam, M. S. (2015), 'A review on natural fiber reinforced polymer composite and its applications', *International Journal of Polymer Science*, **2015**, 1–15.

Murali, G. and Pannirselvam, N. (2011), 'Flexural strengthening of reinforced concrete beams using fibre reinforced polymer laminate: a review', *Journal of Engineering and Applied Sciences*, **6** (11), 41–47.

Nair, L. S. and Laurencin, C. T. (2007), 'Biodegradable polymers as biomaterials', *Progress in Polymer Science*, **32** (9), 762–798.

Noorunnisa Khanam, P., Abdul Khalil, H. P. S., Jawaid, M., Ramachandra Reddy, G., Surya Narayana, C. and Venkata Naidu, S. (2010), 'Sisal/carbon fibre reinforced hybrid composites: tensile, flexural and chemical resistance properties', *Journal of Polymers and the Environment*, **18** (4), 727–733.

Polizu, S., Savadogo, O., Poulin, P. and Yahia, L. (2006), 'Applications of carbon nanotubes-based biomaterials in biomedical nanotechnology', *Journal of Nanoscience and Nanotechnology*, **2006** (7), 1883–1904.

Pothan, L. A., Mai, Y. W., Thomas, S. and Li, R. K. Y. (2008), Tensile and flexural behavior of sisal fabric/polyester textile composites prepared by resin transfer molding technique', *Journal of Reinforced Plastics and Composites*, **27** (16–17), 1847–1866.

Qiao, B., Zhou, D., Dai, Z., Zhao, W., Yang, Q., Xu, Y., Li, X., Wu, J., Guo, S. and Jiang, D. (2019), 'Bone plate composed of a ternary nano hydroxyapatite/polyamide 66/glass fiber composite: biocompatibility in vivo and internal fixation for canine femur fractures', *Advanced Functional Materials* **29** (22), 1808738.

Ramakrishna, S., Mayer, J., Wintermantel, E. and Leong, K.W. (2001), 'Biomedical applications of polymer-composite materials: a review', *Composite Science and Technology*, **61** (9), 1189–1224.

Rao, K. M. M. and Rao, K. M. (2007), 'Extraction and tensile properties of natural fibers: Vakka, date and bamboo', *Composite Structures*, **77** (3), 288–295.

Rezwan, K., Chen, Q. Z., Blaker, J. J. and Boccaccini, A. R. (2006), 'Biodegradable and bioactive porous polymer/inorganic composite scaffolds for bone tissue engineering', *Biomaterials*, **27** (18), 3413–3431.

Samiezadeh, S., Avval, P. T., Fawaz, Z. and Bougherara, H. (2014), 'Biomechanical assessment of composite versus metallic intramedullary nailing system in femoral shaft fractures: A finite element study', *Clinical Biomechanics*, **29** (7), 803–810.

Sarikaya, E., Çallioglu, H. and Demirel, H. (2019), 'Production of epoxy composites reinforced by different natural fibers and their mechanical properties', *Composites Part B: Engineering*, **167**, 461–466.

Sarwar, A., Mahboob, Z., Zdero, R. and Bougherara, H. (2020), 'Mechanical characterization of a new Kevlar/Flax/epoxy hybrid composite in a sandwich structure', *Polymer Testing* **90**, 106680.

Scholz, M. S., Blanchfield, J. P. and Bloom, L.D. (2011), 'The use of composite materials in modern orthopaedic medicine and prosthetic devices: A review', *Composite Science and Technology*, **71** (16), 1791–1803.

Senthilkumar, K., Saba, N., Rajini, N., Chandrasekar, M., Jawaid, M., Suchart Siengchin, and Othman Alotman, Y. (2018), 'Mechanical properties evaluation of sisal fibre reinforced polymer composites: a review', *Construction and Building Materials* **174**, 713–729.

Shibata, S., Cao, Y. and Fukumoto, I. (2008), 'Flexural modulus of the unidirectional and random composites made from biodegradable resin and bamboo and kenaf fibres', *Composites Part A: Applied Science and Manufacturing*, **39** (4), 640–646.

Soundhar, A. and Jayakrishna, K. (2019), 'Investigations on mechanical and morphological characterization of chitosan reinforced polymer nanocomposites', *Materials Research Express* **6** (7), 075301.

Soundhar, A. Jayakrishna, K., Md Shah, A. U., Sultan, M. T. H., Safri, S. N. A. and Mustapha, F. (2020), 'Investigations on the mechanical properties of glass fiber/sisal fiber/chitosan reinforced hybrid polymer sandwich composite scaffolds for bone fracture fixation applications', *Polymers* 12(7), 1501.

Suner, S., Bladen, C. L., Gowland, N., Tipper, J. L. and Emami, N. (2014), 'Investigation of wear and wear particles from a UHMWPE/multi-walled carbon nanotube nanocomposite for total joint replacements', *Wear*, **317** (1–2), 163–169.

Teoh, S.H. (2000), 'Fatigue of biomaterials: a review', *International Journal of Fatigue*, **22** (10), 825–837.

Unterweger, C., Bruggemann, O. and Furs, C. (2014), 'Synthetic fibers and thermoplastic short-fiber-reinforced polymers: properties and characterization', *Polymer Composites*, **35** (2), 227–236.

Valente, M., Sarasini, F., Marra, F., Tirillò, J. and Pulci, G. (2011), 'Hybrid recycled glass fiber/wood flour thermoplastic composites: manufacturing and mechanical characterization', *Composites Part A: Applied Science and Manufacturing* **42** (6), 649–657.

Wolinsky, P., Tejwani, N., Richmond, J. H., Koval, K. J., Egol, K. and Stephen, D. J. (2001), 'Controversies in intramedullary nailing of femoral shaft fractures', *Journal of Bone and Joint Surgery* **83** (9), 1404–1415.

Xie, X., Guo, Z. and Yuan, C. (2020), 'Investigating the water lubrication characteristics of sisal fiber reinforced ultrahigh-molecular-weight polyethylene material', *Polymer Composites* **41** (12), 5269–5280.

Yang, J., Gerry, L. K., Guang, C., Linhui, Z., Antonios, G. M. and Fuzhai, C. A. (2018), 'Review on the exploitation of biodegradable magnesium-based composites for medical applications', *Biomedical Materials*, **13** (2), 022001.

Yang, Y., Boom, R., Irion, B., Heerden, D. J., Kuiper, P. and Wit, H. (2012), 'Recycling of composite materials', *Chemical Engineering and Processing: Process Intensification*, **51**, 53–68.

12 Failure of Polymer Matrix in Space Applications

S. Arulvel
Vellore Institute of Technology (VIT), Vellore, India

D. Dsilva Winfred Rufuss
Vellore Institute of Technology (VIT), Vellore, India
University of Birmingham, UK

Takeshi Akinaga
Akita University, Akita, Japan

CONTENTS

12.1 INTRODUCTION

The main requirements for materials in the aerospace and aircraft industry are low density, fracture resistance, fatigue resistance and damage tolerance. While these requirements have mostly been satisfied by metallic and ceramic materials, their density is not suitable for use in the aerospace and aircraft industry. Researchers have therefore been keen to find an alternate material to ceramic, alloy and metallic materials with a low weight-to-strength ratio.

Polymers are typical material used in various industrial applications owing to their unique characteristics, including high-modulus, strength-to-weight ratio, high toughness, chemical inertness, corrosion resistance, low density, low electrical conductivity, heat conductivity, low cost and easy processing (Bhat and Kandagor, 2014). The unique properties of materials mainly depend on the types of molecules bonded to each other. In this respect, polymers exhibit different properties such as

DOI: 10.1201/9781003128861-12

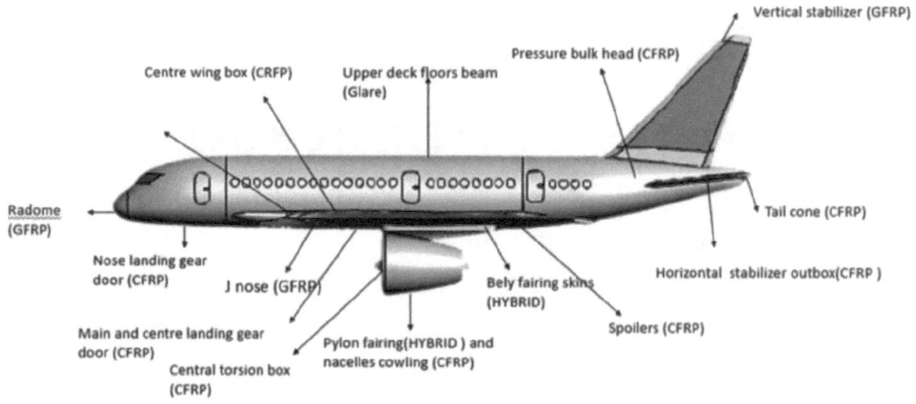

FIGURE 12.1 Applications of PMC in aircraft components.

bending (rubber), stretch (polyester), hardening (epoxy) and toughness (glass), which have extended their application in the aerospace industry (Masuelli, 2013). In addition to this, polymers have been used as additives to join aerospace and aircraft components that are predominantly subject to failure (Figure 12.1).

In terms of structural applications, polymers fail at low-load conditions below their tensile strength due to cyclic stresses and creep rupture. Polymers are therefore not greatly used in structural applications in their existing form, but rather as a composite matrix material with ceramic/fibre/metallic reinforcements. Carbon fibre-reinforced polymer composites are one of the main polymer composites typically used for aerospace structural applications and have been found versatile in all spacecraft and aircraft industries (Rajak et al., 2019). Carbon fibre was first developed from rayon in 1950 and was first introduced for compressor blades by Rolls Royce in 1960. Catastrophic failure of compressor blades occurred due to the high brittleness of carbon fibre. These carbon fibres were subsequently used as reinforcement for polymer matrices to form polymer composites (Kedward, 1997).

The introduction of polymer matrices has certainly increased the toughness and ductility of composites with high mechanical strength. As a result, more than 1,500 composites have been developed as effective replacements for metallic parts used in the aircraft industry (Mouritz, 2012). The availability of different types of fibre (carbon/kevlar/glass/aramid) has made for effective combinations with polymer matrices that had previously produced high-strength polymer composites (Das et al., 2019). However, the challenge for polymer composite researchers is the weakening of the interface strength between polymer matrix and fibre. Due to its high chemical inertness, the polymer matrix does not have an effective interfacial bond with the fibres, which can decrease the tensile strength of polymer composites. Poor interfacial bonding also causes too many mechanical failures, such as fracture, creep rupture, yielding, swelling, impact and crazing, in polymer composites.

Wear failure is one of the significant failures associated with polymer materials in aerospace industry. Different types of polymer matrix composites and polymer coatings are used to restrict wear and erosion in the interior and exterior parts of aircraft, especially in the areas of electromagnetic (EM) shielding, aircraft tyres and the front fan sections of gas turbine engines.

This chapter discusses both mechanical and wear mechanism failures.

12.2 MECHANICAL FAILURES

The application of polymer matrix composites has been extended to aerospace components including frames and seats. They are especially used in the air-flow path of the gas turbine engines (front fan section) due to their specific strength. Although high-temperature material is not required for the fan section of turbine engines, materials with high stiffness, erosion resistance, abrasion resistance and corrosion resistance are mandatory due to the action of strong centrifugal forces during the engine operation.

Polymer matrix composite (PMCs) have generally been used in the fan section of turbine engines. However, their application is limited by poor erosion resistance. Numerous investigations have sought to augment the properties of PMCs through nano additives, where failure mechanisms also predominate. Particular attention to PMC failure mechanisms is thus required in order to develop a innovative PMC materials in the future.

12.2.1 Natural-Fibre PMC

The strength of the matrix plays a vital role in the properties of PMCs. Several polymer matrices (epoxy, phenolics, HDPE, LDPE, acrylics, polypropylene and nylon) were used in the fabrication of composites based on the applications are listed in Tables 12.1 and 12.2. Among the various polymer matrices, epoxy matrix was the thermosetting

TABLE 12.1
Advantages and Disadvantages of Various PMCs

S. No	Types of PMC	Advantages (minimum 5 point)	Disadvantages (minimum 5 point)
1	Ceramic-reinforced PMC	• High strength and hardness • High service temperature • Chemical inertness and rarity	• Low impact resistance • Brittle fracture • Part size and shape limitations • Defect size effect • Limited load level during sliding
2	Polymer-reinforced PMC	• Low operating temperature • High coefficient of thermal and moisture expansion • Low elastic properties in certain directions	• High strength • low cost • High chemical resistance and good insulating property
3	Metal-reinforced PMC	• Higher elastic properties • Higher service temperature • Insensitive to moisture • Higher electric and thermal conductivity • Better fatigue and flow resistance	• Higher processing temperature • Higher densities

TABLE 12.2
Properties of Various Matrices used in PMCs

		Properties					
S. No	Types of Polymer Matrix	Density g/cm³	Melting point °C	Glass transition temperature	Tensile strength MPa	Tensile modulus GPa	References
1	Epoxy	1.15	34	133.74	26	3.1	(Wang et al., 2013) (Bhatia et al., 2019) (Xie et al., 2019)
2	Phenolics	1.07	40.5	218	24.1	2.78	(Kumar et al., 2009)
3	HDPE	.954	135	148	15.7	732.5	(Amjadi and Fatemi, 2020)
4	LDPE	0.91	115	−125	11.8	2.21	(Romisuhani et al., 2010)
5	Acrylics	1.18	160	105	68	3.2	(Kumar et al., 2002)
6	Polypropylene	0.92	160	−20	29	4.6	(Song et al., 2005)
7	Nylon	1.15	256	70	896	5.17	(Song et al., 2005)

polymer used regularly in the fabrication of PMCs. In the early days, natural fibres like cotton, silk, fur, jute, flax, wool and hemp were used as reinforcement fibres in PMCs for many industrial applications. However, these are rarely used for aerospace applications due to the fibre pull-out failure mechanism in high load conditions, which is mainly caused by poor adhesion between matrix and fibre. Though various surface treatments of natural fibres have been used to improve the adhesion between fibre and matrix, their application has been restricted in the aerospace field due to considerations of thermal stability, chemical stability and a number of adverse effects.

12.2.2 Glass-Fibre/Carbon-Fibre PMC

Carbon fibres and glass fibres are routinely used in PMCs for aerospace applications as alternatives to natural fibres, due to itshighly resistant to erosion, and to thermal and cryogenic stresses. The interfacial adhesion between fibre and matrix has been found to be superior to that of natural-fibre PMCs. The major failure mechanisms generally reported for glass fibre-reinforced PMCs are fibre rupture, peeling, delamination and matrix cracking. Research was carried out to improve the mechanical properties of PMCs that demonstrated different failure mechanisms in the past through additives such as Al_2O_3, SiC, SiO_2 and MWCNTs.

In addition to reinforcement, the direction of applied stress can also influence the PMC failure mechanism. Glass fibre-reinforced epoxy tubulars exhibited various failures, including tensile axial failure, distributed weepage, local leakage, burst and compressive axial failure, when subjected to the multiaxial stress (Meijer and Ellyin, 2008). Similarly, bearing failure load on glass-fibre/epoxy (GFRP) laminates demonstrated net-section failure, shear-out failure, bearing failure and fastener shear failure (Ascione et al., 2009). Megahed et al. (2021) substituted an Al mesh layer for the glass-fibre layer. However, there is a decrease in tensile strength as well as flexural strength because of the poor bonding between the adjacent glass fibre and aluminium layers. Soykok et al. (2013) studied glass fibre-reinforced epoxy composites that absorb moisture during immersion in hot water, and found that both the increase in water temperature and the extension of holding time can increase the amount of moisture. Also composite joints treated at a high water temperature suffer a reduction of failure strength. The behaviour of composite joints in tensile loading is affected by holding time in hot water. Long immersion time reduces the strength of mechanically fastened joints. Soykok (2012) investigated the effects of thermal and tightening torque on the failure load on mechanically fastened glass-fibre/epoxy composite joints. In such joints, the load-carrying capacity decreases with increasing temperature and applied torque. A GF/epoxy composite under low-velocity impact loading was investigated by Nassir et al. (2021). The combination of epoxy and glass fibre helps to increase the absorbed energy of the targets. The failure mode and failure load can also be reduced by considering the effect of E/D (edge-to-diameter), S/D (width-to-diameter) and P/D (pitch-to-diameter) ratios when designing multi-pin joints in glass-fibre/epoxy composite laminates (Kishore et al., 2009). Though epoxy is used as a regular polymer matrix, it is easily flammable and releases a substantial amount of smoke and gases. A protective fire-retardant coating is provided around the composite. Geo-polymer material used as fire-retardant material or filler improves the thermal and mechanical properties (in terms of tensile, flexural and impact strength) of glass fibre-reinforced epoxy composite (Shahari et al., 2021). Reinforcement with fire retardant can also pave the way for the use of epoxy in airplanes, land and water vehicles, where a high specific rigidity and strength, high damping, high resistance to corrosion and thermal expansion are required.

A hybrid fibre PMC combination has also been prepared as its flexural strength is greater than glass fibre and carbon fibre-reinforced composite (Turla et al., 2014). The plane strain fracture toughness values for carbon fibre-reinforced polymer composite is greater (53%) than glass fibre-reinforced polymer composite. Natural fibres were also used in the hybrid PMC combination. The combination of natural fibres (*Prosopis juliflora*) with glass fibres has reduced failures caused by matrix cracking, gap creation and low interfacial bonding (Palanivendhan et al., 2021). The *P. juliflora* fills up the gaps in the matrix region and provides more continuity. This uniform distribution allows load transfer between matrix and reinforcement and increases the tensile strength of the composite.

12.3 WEAR FAILURE

Wear failure is one of the most significant restrictions on the application of PMC in aircraft. The strength of the interface between the matrix and particle are vital to both mechanical and tribology properties (Wan et al., 2006). Due to thermal stability, the use of PMC in several aero components has been restricted. However, it is unavoidable in those regions of aircraft not subject to high temperatures. For example, the application of PMC can be seen in the inlet sections of engine components. The highly erosion-resistant PMC material is mostly preferred at the engine inlets of the aircraft due to the entry of dust and erosive particles. Researchers have improved the wear resistance of PMC through various reinforcements and fibres.

Carbon-fibre/epoxy matrix composite, in which both matrix and fibre are subject to slow wear, is one of the strong alternatives to glass-fibre/epoxy composite (Schön, 2004). The major wear mechanisms observed for the PMC are fibre/particle pull-out, matrix degradation, micro grooves (Lee and Hwang, 2006) and formation of brittle cracks at the interfaces. A significant part of the load applied to a joint at quasi-static failure will be transferred by friction force; hence, it is important to reduce the friction force to limit failures in the carbon-fibre/epoxy matrix composite. Friction force can be reduced by improving the thermal and mechanical properties of the carbon-fibre/epoxy matrix composite. The fibre content (Khan et al., 2020) and various treatments like air oxidation, cryogenic (Zhang and Zhong, 2004), alkaline (Verma et al., 2021) and saline treatment (Cervantes et al., 2020) have improved the properties of the carbon-fibre /epoxy matrix composite. Longer carbon-fibre composites showed better wear resistance than shorter carbon-fibre composites (Zhang et al., 2007). However, the wear resistance of shorter-fibre carbon-fibre composites was better with reinforcements like graphite, polytetrafluoroethylene (PTFE) and TiO_2 (Zhang et al., 2004; Mesbahi et al., 2012).

Milled carbon fibre-reinforced epoxy gradient composites were also developed at different centrifugal speeds and gave the best wear resistance (Chand and Sharma, 2008). In addition to the fibre length and treatments, the orientation of the fibre also reduced the friction and wear rate of PMCs (Dhieb et al. 2016). For unidirectional fibres, it was fibre de-bonding wear that was the main failure mechanism for carbon-fibre/epoxy composites. The glass-fibre/epoxy composite reinforced with Al_2O_3 nanoparticles exhibited a higher erosion resistance than the glass-fibre/epoxy composite reinforced with micro-powders (Ahmedizat et al., 2020). However, reinforcements such as carbon nanotubes (CNTs) and graphite showed no improvement in the wear resistance of carbon-fibre/epoxy composites (Papadopoulos et al., 2016; Rao et al., 2015). It is also observed that fibre breakage and fibre matrix de-bonding was lower for unfilled composites than for graphite-filled composites. Multi-walled carbon nanotubes reinforced with E-glass fibre increased the impact strength, tensile strength and hardness value compared to other varieties of epoxy hybrid composites (Nadeem et al., 2021). Reinforcement particles strongly affect the mechanical and tribology properties of glass fibre, whereas their effect on carbon fibre was limited. This could be due to poor interfacial adhesion between the carbon fibre, matrix and reinforcement.

Hybridization can improve the wear resistance of composite fibres including glass/carbon/natural fibres (Velu et al., 2021; Suresha et al., 2007; Sarkar et al., 2017). The wear rate of glass fibre is comparatively lesser than that of carbon fibre.

12.4 SUMMARY

The following conclusions can be drawn pertaining to PMC:

- The mechanical and tribology strength of carbon fibre was comparatively better than that of PMCs reinforced with natural fibre and glass fibre. The majority of wear mechanisms reported for the PMCs are fibre/particle pull-out and matrix degradation.
- For several PMCs, the cause of fibre de-bonding is reported to be due to poor interfacial adhesion between the matrix, fibres and reinforcement fibres.
- Polymer matrix failure is due to load transfer directly to the matrix rather than to the fibres. Innovative reinforcements are required to enhance the damping resistance of the matrix so as to reduce brittle failure.
- Reinforcement particles (MWCNT, Al_2O_3, PTFE and TiO_2) play a vital role in improving the mechanical and wear resistance of glass fibre-reinforced PMCs. However, carbon fibre-reinforced PMCs are less influential.
- Hybrid combinations of fibres demonstrate better properties than reinforced PMCs.
- Production of highly erosion and corrosion-resistant PMCs are especially required for aerospace applications. This should hence be a focus for future research.

REFERENCES

Ahmedizat, Shatha Riyad, Aseel Basim Al-Zubaidi, Ahmed A. Al-Tabbakh, Amine Achour, and Alaa Abdul Hamead. Comparative study of erosion wear of glass fiber/epoxy composite reinforced with Al_2O_3 nano and micro particles. *Materials Today: Proceedings* 20 (2020): 420–427.

Ascione, Francesco, Luciano Feo, and Franco Maceri. An experimental investigation on the bearing failure load of glass fibre/epoxy laminates. *Composites Part B: Engineering* 40, no. 3 (2009): 197–205.

Amjadi, Mohammad, and Ali Fatemi. Tensile behavior of high-density polyethylene including the effects of processing technique, thickness, temperature, and strain rate. *Polymers* 12, no. 9 (2020): 1857.

Bhatia, Sunny, Surjit Angra, and Sabah Khan. Mechanical and wear properties of epoxy matrix composite reinforced with varying ratios of solid glass microspheres. In *Journal of Physics: Conference Series*, IOP Publishing, UK 2019, Vol. 1240, no. 1, p. 012080.

Bhat, G. and V. Kandagor. Synthetic polymer fibers and their processing requirements. In *Advances in Filament Yarn Spinning of Textiles and Polymers*, Woodhead Publishing, UK, 2014, pp. 3–30.

Chand, Navin and Manoj K. Sharma. Development and sliding wear behaviour of milled carbon fibre reinforced epoxy gradient composites. *Wear* 264, no. 1–2 (2008): 69–74.

Das, Tushar Kanti, Prosenjit Ghosh, and Narayan Ch Das. Preparation, development, outcomes, and application versatility of carbon fiber-based polymer composites: a review. *Advanced Composites and Hybrid Materials* (2019): 1–20.

Dhieb, H., J.G. Buijnsters, K. Elleuch, and J.-P. Celis. Effect of relative humidity and full immersion in water on friction, wear and debonding of unidirectional carbon fiber reinforced epoxy under reciprocating sliding. *Composites Part B: Engineering* 88 (2016): 240–252.

Dzul-Cervantes, M.A.A., O.F. Pacheco-Salazar, L.A. Can-Herrera, M.V. Moreno-Chulim, J.I. Cauich-Cupul, P.J. Herrera-Franco, and A. Valadez-González. Effect of moisture content and carbon fiber surface treatments on the interfacial shear strength of a thermoplastic-modified epoxy resin composites. *Journal of Materials Research and Technology* 9, no. 6 (2020): 15739–15749.

Kedward, Keith T. Large composite fan blade development for modern aeroengines. In *Proc. 11th Int. Conf. on Composite Materials*, 1 (1997): 200–217.

Khan, Zahid Iqbal, Agus Arsad, Zurina Mohamad, Unsia Habib, and Muhammad Abbas Ahmad Zaini. Comparative study on the enhancement of thermo-mechanical properties of carbon fiber and glass fiber reinforced epoxy composites. *Materials Today: Proceedings 39* (2020) 956–958.

Kishore, A. Nanda, S. K. Malhotra, and N. Siva Prasad. Failure analysis of multi-pin joints in glass fibre/epoxy composite laminates. *Composite Structures* 91, no. 3 (2009): 266–277.

Kumar, N. Mithil, G. Venkata Reddy, S. Venkata Naidu, T. Shobha Rani, and M.C.S. Subha. Mechanical properties of coir/glass fiber phenolic resin based composites. *Journal of Reinforced Plastics and Composites* 28, no. 21 (2009): 2605–2613.

Kumar, Satish, Harit Doshi, Mohan Srinivasarao, Jung O. Park, and David A. Schiraldi. Fibers from polypropylene/nano carbon fiber composites. *Polymer* 43, no. 5 (2002): 1701–1703.

Lee, Hak Gu and Hui Yun Hwang. Effect of wear debris on the tribological characteristics of carbon fiber epoxy composites. *Wear* 261, no. 3–4 (2006): 453–459.

Masuelli, Martin Alberto. Introduction of fibre-reinforced polymers– polymers and composites: concepts, properties and processes. In *Fiber Reinforced Polymers-the Technology Applied for Concrete Repair*. IntechOpen, UK, 2013.

Megahed, A.A., F. Abd El-Wadoud, A. Wagih, and A.M. Kabeel. Effect of incorporating aluminum wire mesh on the notched and un-notched strengths of glass fiber/epoxy composites. *Composite Structures* (2021): 113695.

Meijer, Garret, and Fernand Ellyin. A failure envelope for ±60 filament wound glass fibre reinforced epoxy tubulars. *Composites Part A: Applied Science and Manufacturing* 39, no. 3 (2008): 555–564.

Mesbahi, Ali Haghighat, Dariush Semnani, and Saeid Nouri Khorasani. Performance prediction of a specific wear rate in epoxy nanocomposites with various composition content of polytetrafluoroethylen (PTFE), graphite, short carbon fibers (CF) and nano-TiO2 using adaptive neuro-fuzzy inference system (ANFIS). *Composites Part B: Engineering* 43, no. 2 (2012): 549–558.

Mouritz, Adrian P. *Introduction to Aerospace Materials*. Elsevier, Woodhead Publishing Limited, UK, 2012.

Nadeem, M., S. Chethan, K. Srinivasa, M. Karthik Kumar, and N. Yathisha. Effect of glass fibers and multi walled carbon nano tubes (MWCNT's) on mechanical properties of epoxy hybrid composites at elevated temperature. *Materials Today: Proceedings* 44 (2021) 2013–2018.

Nassir, Nassier A., R. S. Birch, W. J. Cantwell, M. Al Teneiji, and Z. W. Guan. The Perforation Resistance of Aluminum-Based Thermoplastic FMLs. *Applied Composite Materials* 28 (2021): 587–605.

Palanivendhan M., J. Chandradass, T. Kaviyarasu, and J. Philip. Fabrication and characteristics of hybrid glass fiber/Prosopis Juliflora reinforced epoxy composite. *Materials Today: Proceedings* 45 (2021): 6833–6837.

Papadopoulos, A., G. Gkikas, A.S. Paipetis, and N.-M. Barkoula. Effect of CNTs addition on the erosive wear response of epoxy resin and carbon fibre composites. *Composites Part A: Applied Science and Manufacturing* 84 (2016): 299–307.

Rajak, Dipen Kumar, Durgesh D. Pagar, Pradeep L. Menezes, and Emanoil Linul. Fiber-reinforced polymer composites: manufacturing, properties, and applications. *Polymers* 11, no. 10 (2019): 1667.

Rao, K.S., Y.S. Varadarajan, and N. Rajendra. Erosive wear behaviour of carbon fiber-reinforced epoxy composite. *Materials Today: Proceedings* 2, no. 4–5 (2015): 2975–2983

Romisuhani, A., H. Salmah, and H. Akmal. Tensile properties of low density polypropylene (LDPE)/palm kernel shell (PKS) biocomposites: the effect of acrylic acid (AA). In *IOP Conference Series: Materials Science and Engineering*, IOP Publishing, 2010, vol. 11, no. 1, p. 012001.

Sarkar, Pujan, Nipu Modak, and Prasanta Sahoo. Effect of normal load and velocity on continuous sliding friction and wear behavior of woven glass fiber reinforced epoxy composite. *Materials Today: Proceedings* 4, no. 2 (2017): 3082–3092.

Schön, Joakim. Coefficient of friction and wear of a carbon fiber epoxy matrix composite. *Wear* 257, no. 3–4 (2004): 395–407.

Shahari, Shazzuan, M. Fathullah, Mohd Mustafa Al Bakri Abdullah, Z. Shayfull, Mozammel Mia, and Vertic Eridani Budi Darmawan. Recent developments in fire retardant glass fibre reinforced epoxy composite and geopolymer as a potential fire-retardant material: a review. *Construction and Building Materials* 277 (2021): 122246.

Song, P.S., S. Hwang, and B.C. Sheu. Strength properties of nylon-and polypropylene-fiber-reinforced concretes. *Cement and Concrete Research* 35, no. 8 (2005): 1546–1550.

Soykok, Ibrahim Fadil, Onur Sayman, Mustafa Ozen, and Behiye Korkmaz. Failure analysis of mechanically fastened glass fiber/epoxy composite joints under thermal effects. *Composites Part B: Engineering* 45, no. 1 (2013): 192–199.

Soykok, İbrahim Fadıl. *Failure analysis of bolted and pinned composite joints under temperature effects*. PhD diss., DEÜ Fen Bilimleri Enstitüsü (2012).

Suresha, B., G. Chandramohan, P. Samapthkumaran, and S. Seetharamu. Three-body abrasive wear behaviour of carbon and glass fiber reinforced epoxy composites. *Materials Science and Engineering: A* 443, no. 1–2 (2007): 285–291.

Turla, Prashanth, S. Sampath Kumar, P. Harshitha Reddy, and K. Chandra Shekar. Processing and flexural strength of carbon fiber and glass fiber reinforced epoxy-matrix hybrid composite. *International Journal of Engineering Research and Technology (IRJET)*, 3 (2014).

Velu, S., J.K. Joseph, M. Sivakumar, V.K. Bupesh Raja, K. Palanikumar, and N. Lenin. Experimental investigation on the mechanical properties of carbon-glass-jute fiber reinforced epoxy hybrid composites. *Materials Today: Proceedings* 46 (2021) 3566–3571.

Verma, Rajneesh, Mukul Shukla, and Dharmendra Kumar Shukla. Effect of glass fibre hybridization on the water absorption and thickness of alkali treated kenaf-epoxy composites. *Materials Today: Proceedings* 44 (2021) 2093–2096.

Wan, Y.Z., G.C. Chen, S. Raman, J.Y. Xin, Q.Y. Li, Y. Huang, Y.L. Wang, and H.L. Luo. Friction and wear behavior of three-dimensional braided carbon fiber/epoxy composites under dry sliding conditions. *Wear* 260, no. 9–10 (2006): 933–941.

Wang, Zhenqing, Fang Liu, Wenyan Liang, and Limin Zhou. Study on tensile properties of nanoreinforced epoxy polymer: macroscopic experiments and nanoscale FEM simulation prediction. *Advances in Materials Science and Engineering* 2013 (2013).

Xie, Zongliang, Hongliang Zhang, Siyu Zhang, He Li, Xialin Su, Peng Liu, and Zongren Peng. Effect of moisture on electrical properties of epoxy/paper composites. In *2019 IEEE Conference on Electrical Insulation and Dielectric Phenomena (CEIDP)*, IEEE, 2019, pp. 174–177.

Zhang, Hui and Zhong Zhang. Comparison of short carbon fibre surface treatments on epoxy composites: II. Enhancement of the wear resistance. *Composites Science and Technology* 64, no. 13–14 (2004): 2031–2038.

Zhang, Z., C. Breidt, L. Chang, F. Haupert, and K. Friedrich. Enhancement of the wear resistance of epoxy: short carbon fibre, graphite, PTFE and nano-TiO_2. *Composites Part A: Applied Science and Manufacturing* 35, no. 12 (2004): 1385–1392.

Zhang, Hui, Zhong Zhang, and Klaus Friedrich. Effect of fiber length on the wear resistance of short carbon fiber reinforced epoxy composites. *Composites Science and Technology* 67, no. 2 (2007): 222–230.

Index

O

P

R

S

T

U

V

W

Y